WITHDRAWN
UTSA LIBRARIES

Managing and Leading

RENEWALS 458-4574

Also of Interest

▶ *Guide to Hiring and Retaining Great Civil Engineers* (ASCE Manuals and Reports on Engineering Practice No. 103). 2003. ISBN: 0-7844-0627-8. Price: $29.00.

▶ *How to Work Effectively with Consulting Engineers: Getting the Best Project at the Right Price* (ASCE Manuals and Reports on Engineering Practice No. 45, revised edition). 2003. ISBN: 0-7844-0637-5. Price: $29.00.

▶ *Ten Commandments of Better Contracting: A Practical Guide to Adding Value to an Enterprise through More Effective SMART Contracting*, Francis T. Hartman. 2003. ISBN: 0-7844-0653-7. Price: $79.00.

▶ *Civil Engineering Practice in the Twenty-First Century: Knowledge and Skills for Design and Management*, Neil S. Grigg, Thomas J. Siller, Darrell G. Fontane, and Marvin E. Criswell. 2001. ISBN: 0-7844-0526-3. Price: $39.00.

▶ *Collective Excellence: Building Effective Teams, 2nd Edition*, Mel Hensey. 2001. ISBN: 0-7844-0546-8. Price: $24.00.

▶ *Diversity: A Special Issue of Leadership and Management in Engineering*, edited by Jeffrey S. Russell. 2001. ISBN: 0-7844-0588-3. Price: $15.00.

▶ *Engineering Your Future, Second Edition: The Non-Technical Side of Professional Practice in Engineering and Other Technical Fields*, Stuart G. Walesh. 2000. ISBN: 0-7844-0489-5. Price: $45.00.

▶ *Personal Success Strategies: Developing Your Potential*, Mel Hensey. 1999. ISBN: 0-7844-0446-1. Price: $25.00.

▶ *Continuous Excellence: Building Effective Organizations*, Mel Hensey. 1995. ISBN: 0-7844-0013-X. Price: $23.00.

Managing and Leading

52 Lessons Learned for Engineers

Stuart G. Walesh, Ph.D, P.E.

Library
University of Texas
at San Antonio

Library of Congress Cataloging-in-Publication Data
Walesh S.G.
  Managing and Leading : 52 lessons learned for engineers / Stuart G. Walesh.
  p. cm.
  Includes bibliographical references and index.
  ISBN 0-7844-0675-8
  1. Engineering—Management I. Title
  TA190.W34 2003
  620'.0068—dc22   2003062776

Published by American Society of Civil Engineers
1801 Alexander Bell Drive
Reston, Virginia 20191
www.asce pubs.asce.org

Any statements expressed in these materials are those of the individual authors and do not necessarily represent the views of ASCE, which takes no responsibility for any statement made herein. No reference made in this publication to any specific method, product, process or service constitutes or implies an endorsement, recommendation, or warranty thereof by ASCE. The materials are for general information only and do not represent a standard of ASCE, nor are they intended as a reference in purchase specifications, contracts, regulations, statutes, or any other legal document. ASCE makes no representation or warranty of any kind, whether express or implied, concerning the accuracy, completeness, suitability, or utility of any information, apparatus, product, or process discussed in this publication, and assumes no liability therefore. This information should not be used without first securing competent advice with respect to its suitability for any general or specific application. Anyone utilizing this information assumes all liability arising from such use, including but not limited to infringement of any patent or patents.

ASCE and American Society of Civil Engineers—Registered in U.S. Patent and Trademark Office.

Photocopies: Authorization to photocopy material for internal or personal use under circumstances not falling within the fair use provisions of the Copyright Act is granted by ASCE to libraries and other users registered with the Copyright Clearance Center (CCC) Transactional Reporting Service, provided that the base fee of $18.00 per article is paid directly to CCC, 222 Rosewood Drive, Danvers, MA 01923. The identification for ASCE Books is 0-7844-0675-8/04/ $18.00. Requests for special permission or bulk copying should be addressed to Permissions & Copyright Dept., ASCE.

Copyright © 2004 by the American Society of Civil Engineers.
All Rights Reserved.
Library of Congress Catalog Card No: 2003062776
ISBN 0-7844-0675-8
Manufactured in the United States of America.

Library
University of Texas
at San Antonio

Dedication

To my daughter, Kim,
who shares a love of ideas and the
challenge of finding ways to share them.

Contents

Abbreviations and Acronyms

A/E	architect/engineer
ABET	Accreditation Board for Engineering and Technology
ACEC	American Council of Engineering Companies
AH HA!	Awareness-Head-Heart-Action
AMA	American Management Association
ASCE	American Society of Civil Engineers
ASQC	American Society of Quality Control
AOM	Academy of Management
ASME	American Society of Mechanical Engineers
BOK	Body of Knowledge
CADD	computer-aided drafting and design
CCL	Center for Creative Leadership
CEO	Chief Executive Officer
CEU	continuing education unit
DAD	decide-announce-defend
DWYSYWD	do what you said you would do
E&T	education and training
IEEE	Institute of Electrical and Electronics Engineers
IT	information technology
NASE	National Association for the Self-Employed
NSA	National Speakers Association
NSPE	National Society of Professional Engineers
P^5	preparing, presenting, and publishing professional papers
P.E.	professional engineer
PEEP	Professional Engineers in Professional Practice (of NSPE)
PMI	Project Management Institute
PMP	Project Management Professional
POP	public owns project
ROI	return on investment
RFP	Request for Proposal
RSV	Revised Standard Version (*The Bible*)

SBA	Small Business Administration
SMART	Specific, Measurable, Achievable, Relevant, Time-framed
SOHO	Small Office/Home Office
T^3	Tell them what you are going to tell them. Tell them. Tell them what you told them.
TC	task committee
TEAM	together everyone achieves more
TQM	total quality management
TVA	Tennessee Valley Authority

Preface

*I*n the practice of engineering, technical competency is not sufficient for personal success and fulfillment in consulting, industry, academia, and government. Engineers must improve their managing and leading effectiveness by augmenting technical "hard-side" competencies with complementary "soft-side" knowledge, skills, and attitudes.

Managing and Leading: 52 Lessons Learned for Engineers is designed to help engineers as they carry out their assignments, whether delegated by others or self-assigned. We engineers seem to be under ever-increasing pressure to increase productivity, to accomplish more with what we have and, sometimes, to do more with less. Many of us aspire to be better stewards of our time and talent gifts. We yearn for more than success; we seek significance in what we do. We want to get beyond ourselves, positively affect others, and make a difference. This requires management and leadership ability.

The 52 lessons in this book present useful ideas for ways to more effectively approach the non-technical, soft-side aspects of working with colleagues, clients, customers, the public, and other stakeholders. Each lesson contains an essay that offers at least one idea or principle for honing management and leadership effectiveness. Following each essay are pragmatic suggestions for ways to apply the ideas using application tools and techniques such as action items, guidelines, do's and don'ts, checklists, forms, and resource materials such as articles, papers, books, e-newsletters, and websites. The goal is to help you make better use of who you are and what and who you know.

As a writer and speaker, I find that quotes are an effective way to reinforce or repeat a message. Throughout the book, supportive quotes by notable leaders, writers, and speakers from a number of disciplines help emphasize the global scope and suggest the timeless nature of insightful thoughts on managing, leading, and related matters. A list of quoted individuals is provided in the Appendix.

Many approaches to using this book are possible, the most obvious of which is to consecutively read all 52 lessons in one or more sittings. A more focused approach is to select a lesson or group of lessons from the Contents or the Index that resonates with your current need. Another way, as suggested by the number of lessons, is to select one lesson each week and work through the lessons over one year.

Upon reading a lesson, you may determine its message has potential value for you. If so, commit to putting the underlying ideas or principles into practice, at least on a trial basis, and experiment with some of the suggestions. Perhaps you will leverage the lesson into a new, life-long management and leadership habit. Even if you don't change your method of working, you will have examined and confirmed it on its merits.

Counseling, teaching, new employee orientation, and mentoring are other possible uses of *Managing and Leading.* High school counselors could loan this book to potential engineering students to broaden their understanding of what engineers need to know how to do. It also could be used as a supplemental text for engineering courses that introduce or stress personal and professional development of students. Individual students could be asked to review selected lessons and offer their views. Engineering employers could include a personal or loan copy in the orientation package given to new personnel. Finally, mentors might assign their mentees one or more related lessons for self-study and discussion.

Writing this book has been, in a sense, a personal journey. Most of the essays and supporting applications reflect my positive and negative experiences as I tried first to manage and later to lead in the public, private, and academic sectors. Bracketed by the "ecstasy of victory" and the "agony of defeat," my spectrum of management and leadership experiences prompted me, in the mid-1990s, to begin writing essays about them as a means of reflecting and learning.

Many of the essays, or earlier versions of them, have been individually published or presented elsewhere. Working on this book offered the rare opportunity to reflect further on each essay, refine some, write new ones, add application tools and techniques, and organize 52 of them into an integrated whole. All of the lessons are intended to help individual engineers and some offer guidance to private and public engineering organizations. I hope that some of what I have learned will increase your managing and leading capabilities. No matter how your management or leadership efforts turn out, reflect on and learn from them throughout your career.

Acknowledgments

*I*n addition to the many references and other sources included in this book, I received a wealth of ideas and information from and have been influenced by former students, seminar and workshop participants, clients, colleagues, and friends. Their assistance, intended and otherwise, is sincerely appreciated. In addition, numerous individuals provided assistance during the writing of this book when I was searching for ideas, information, and applications or needed a second opinion on content or style. I thankfully acknowledge the help of Cathy Crossett Avila, P.E.; Ernie Avila, P.E.; Arnold Bandstra, P.E.; William N. Bickel, P.E.; Robert Boller; Dee and John Caruso; Reverend Jerry Castleman; Irene Dorsey; John V. Farr, Ph.D., P.E.; John A. Hardwick, P.E.; William Hayden, Jr., P.E.; Mel Hensey, P.E.; Carol Householder; Dorothy Leeds; Thomas A. Lenox, Ph.D.; Carol and William Minor; Peter L. Monkmeyer, Ph.D., P.E.; Robert Piekarz; Mark Rauckhorst, P.E.; Jeffrey S. Russell, Ph.D., P.E.; James H. Shonk; Jill Walls and Richard G. Weingardt, P.E.

The contributions of members of the ASCE Press team are appreciated. Jack Bruggeman served as the initial Acquisitions Editor, and Bernadette Capelle, his successor, provided overall guidance throughout the project. Book production was capably managed by Suzanne Coladonato, copy editor Lori Peterson added significant value by improving the effectiveness of my writing, and publicity coordinator Allison Stathos creatively led the marketing effort. Finally, Jerrie, my wife, typed and retyped many drafts; meticulously proofed punctuation, spelling, and grammar; critiqued content; and, as always, provided total support.

Stuart G. Walesh, Ph.D., P.E.
Cape Haze, Florida
November 2003

About the Author

*D*r. Stuart G. Walesh, P.E., has more than 40 years of engineering, education, and management experience in the government and private sectors. He earned a B.S. in Civil Engineering at Valparaiso University, an MSE at The Johns Hopkins University, and a Ph.D. from the University of Wisconsin-Madison. He has functioned as a project manager, department head, discipline manager, marketer, professor, and dean of an engineering college.

Walesh's technical specialty is water resources engineering. He has led or participated in watershed planning, computer modeling, flood control, stormwater and floodplain management, groundwater, dam, and lake projects. Other experience includes research and development, engineering economics, stakeholder participation, forensics, and expert analysis and witness services. Areas in which he has provided management and leadership services include strategic and business planning, TQM, reengineering, technical and non-technical education and training, corporate universities, writing and editing, marketing, facilitation, mentoring, coaching, and team building.

As a consultant, Dr. Walesh provides management, engineering, education/training, and marketing services. Clients include the American Society of Civil Engineers, Bonar Group, Boston Society of Civil Engineers, BSA Life Structures, Camp Dresser and McKee, Clark Dietz, Daimler Chrysler, Donohue & Associates, Earth Tech, Hinshaw & Culbertson, Indiana Department of Natural Resources, J.F. New & Associates, MSA Professional Services, PBS&J, Rust Environment & Infrastructure, Taylor Associates, U.S. Environmental Protection Agency, Wright Water Engineers, and the communities of Pendleton and Valparaiso, Indiana.

Walesh is a member of ASCE and NSPE. He has served as member and chair of state and national committees and groups. For example, he served on the ASCE Task Committee on the First Professional Degree, was Special Issues Editor for ASCE's Committee on Publications, and is Vice-Chair of the ASCE Task Committee on Academic Prerequisites for Professional Practice. Walesh served on the Indiana Board of Registration for Professional Engineers. In 1995, he received the Public Service Award from the Consulting Engineers of Indiana; in 1998, the Distinguished Service Citation from the College of Engineering at the University of Wisconsin; and in 2003, the Excellence in Civil Engineering Education Leadership Award presented by the ASCE Educational Activities Committee.

Walesh authored *Urban Surface Water Management* (Wiley, 1989), *Engineering Your Future,* Second Edition (ASCE Press, 2000), *Flying Solo: How to Start an Individual Practitioner Consulting Business* (Hannah Publishing, 2000), and *Managing & Leading: 52 Lessons for Engineers* (ASCE Press, 2004). Walesh is author or co-author of more than 200 presentations, papers, and articles in the areas of engineering, education, and management, and he has facilitated or presented more than 150 workshops, seminars, and meetings throughout the United States.

Personal Roles, Goals, and Development

"Get your ducks in a row" has always been one of my favorite expressions. Another is "begin with the end in mind." These expressions emphasize the importance of preparation, and, while they apply to both groups and individuals, the latter is relevant here.

If we aspire to manage and lead individuals, projects, organizations, and change, we must first get our personal "ducks" in a row. This requires selecting roles, setting goals, and acquiring and developing supportive knowledge, skills, and attitudes that have proven effective. The purpose of the 13 lessons in Part 1 is to help you get your personal act together by thinking about and embracing that knowledge-skills-attitudes set.

1

Leading, Managing, and Producing

Leaders are people who do the right things;
managers are people who do things right.
Both roles are crucial, but they differ profoundly.

Warren G. Bennis

One model for an organization, such as an engineering consulting firm, a manufacturing business, an academic department or college, or a government agency, is that wholeness, vitality, and resiliency require attention to three different, but inextricably related, ongoing functions: leading, managing, and producing (Walesh, 2000). Another way of looking at leading, managing, and producing is to think of them as the three D's: deciding, directing, and doing.

The three different, but complementary, efforts essential to an organization's success may be further explained as follows:

▶ Leading involves deciding what ought to be done or determining where an organization should go. When we are in a leading or deciding mode, helpful knowledge and skills include visioning, communication, honesty and integrity, goal setting and related strategizing, continuous learning, courage, calmness in crises, tolerance for ambiguity, and creativity.

▶ Managing focuses on directing who is doing what when. When we are in a managing or directing mode, useful knowledge and skills include communication, delegation of authority, planning, resource acquisition and allocation, and monitoring.

Leadership works
through people
and culture.
It's soft and hot.
Management works
through hierarchy
and systems.
It's harder and cooler.

John P. Kotter

▶ Producing involves doing what has been decided as a result of leading and what is being directed via managing. When we are in a producing or doing mode, helpful knowledge and skills include technical competence, focus, persistence, and teamwork.

The metaphor of a three-legged stool suggests how use of the leading, managing, and producing model creates a stable organization, one that cannot easily be "knocked over." While an organization or group might temporarily survive balanced on two of the three legs, all three legs are needed for long-term survival. For example, a leaderless manufacturing firm might do well for several years by balancing on the two legs of excellent management and production capabilities, but eventually it will topple because it lacks the third leg. That leg is leadership, especially the ability to see and act on changes in client and customer needs and the means to serve those needs. Leadership author Warren G. Bennis (1989) says:

> Many an institution is very well-managed and very poorly led. It may excel in the ability to handle each day all the routine inputs, yet may never ask whether the routine should be done at all.

Consider another example of the need for three strong legs, with each carrying its share of the weight. A consulting engineering firm, led by a visionary and staffed with individuals willing to produce, lacks the managing leg, that is, effective project managers. This firm is likely to fail because it lacks the ability to translate the vision to profitable services and deliverables.

Assuming you agree that each organization or group striving to be successful has leading (deciding), managing (directing), and producing (doing) responsibilities, consider the manner in which these corporate responsibilities might be met. More specifically, consider the matter of individual responsibility in achieving the three organizational responsibilities.

In what might be called the traditional segregated model, the three functions are performed by three separate groups of personnel. The vast majority of employees or group members are the doers or producers, a distinctly different and much smaller group of managers are the directors, and one person, or perhaps a very small group, leads.

Another traditional way of viewing the production, management, and leadership functions is the linear model. An aspiring and successful individual begins in a production mode and then passes linearly through management and into leadership. Rather than being a trait that many can possess, albeit to different degrees, leadership is considered the end of the line or ultimate destination for a very few. But is this the optimum way for the modern or future organization or group to meet its leading, managing, and producing responsibilities? Probably not.

An organization will be stronger if what used to be the three organizational responsibilities now also become individual responsibilities. The goal should be to enable each member of the organization or group to be a decider, a director, and a doer. While the relative "amounts" of leading, managing, and producing will vary markedly among individuals in the organization or group, everyone should be expected and enabled to do all three in accordance with their individual characteristics.

This shared responsibility organizational model, in contrast with the traditional segregated model, is much more likely to tap, draw on, and benefit from the diverse aspirations, talents, and skills that should be present within the organization or group. Because essentially all members are fully involved, the organization that shares responsibility is in a much better position to synergistically build on internal strengths, to cooperatively diminish internal weaknesses, and to learn about and be prepared to respond to external threats and opportunities.

> *If you want to build a ship, don't drum up the people to gather wood, divide up the work and give orders. Instead, teach them to yearn for the vast and endless sea.*
>
> Antoine de Saint-Exupery

In conclusion, leading, managing, and producing are not defined only or even primarily by position. Instead, the principal determinant is individual knowledge, skills, and attitudes. Anyone at any level within an organization or group can and should exercise deciding, directing, and doing as needs and opportunities arise.

Suggestions for Applying Ideas

Review the examples in the following table for more insight into the distinctions among leading, managing, and producing (or deciding, directing, and doing)

Leading	Managing	Producing
Decide what ought to be done	Direct how things will be done, who will do them, and when	Do what we know has to be done or what we are asked to do
What do we want to accomplish?	How can we best accomplish it?	Do it
Determine if the ladder is leaning against the right wall (Covey, 1990)	Determine how to efficiently climb the ladder (Covey, 1990)	Climb the ladder and go over the wall (Covey, 1990)
Create something new (Kanter, 1993)	Take care of what exists (Kanter, 1993)	Do what needs to be done (Kanter, 1993)
Select jungle to conquer (Covey, 1990)	Sharpen machetes, write policy and procedure manuals, establish working schedules (Covey, 1990)	Cut through the jungle (Covey, 1990)

Look for and act on leadership events

▶ Engineering professor James Parkin (1997) defines a leadership event as "some situation in organizational life, which contains for an individual, an unfilled need for leadership."

▶ Examples of possible leadership events are a failing process within a government office, need to develop a new service in a consulting firm, an increasingly irrelevant curriculum in a university department, and planning a new type of employee gathering within an organization.

> *Nothing is orderly until man takes hold of it. Everything in creation lies around loose.*
>
> Henry Ward Beecher

▶ According to Parkin (1997), the "ability of a person to choose her or his source of event meaning is an important concept as it opens to the potential leader the possibility of changing other actors' event meanings." The point: There are many ways to look at situations that range from—at one end of the spectrum—mundane, routine, and more of the same, to—at the other end of the spectrum—a unique opportunity for improvement and achievement.

▶ Having identified a potential leadership event, move, as appropriate, through the following seven-step process (Parkin, 1997, with parenthetic comments added):

- Confirm that your leadership is required and feasible.
- Identify the potential network of interested actors (determine the stakeholders).
- Estimate the event meanings for the other actors (the "costs" and "benefits" that might be incurred, or perceived to be incurred, by each stakeholder).
- Influence the other actors to share your event meaning and endorse your leadership (engage some in one-on-one idea exchanging conversations).
- Sustain the actor network by communicating and reinforcing roles.
- Organize the actor network to achieve common goals.
- Recognize when goals have been achieved (and celebrate!).

▶ Experience strongly indicates that each of us is surrounded by leadership events, by opportunities to lead regardless of our formal position.

Think about the do's of effective managing (Coe, 1987)

▶ Develop individual and group goals

▶ Communicate objectives

▶ Involve stakeholders

▶ Hold individuals accountable

▶ Put yourself "in their shoes"

▶ Exhibit enthusiasm

▶ Delegate

▶ Keep supervisor informed

▶ Create and consider alternatives

▶ Listen for real meanings

▶ Reflect on the big picture

▶ Prepare for and follow up on meetings

▶ Solicit multidiscipline, including non-technical, input

▶ Get the facts, but accept ambiguity

▶ Accept policies, except when major flaws occur, and then advocate change

▶ Define roles and functions of individuals

▶ Monitor schedules, budgets, and quality

▶ Get out of the office

Manage from the left (brain); lead from the right.
Stephen R. Covey

▶ Provide clear instructions with emphasis on expected results

▶ Participate actively in professional and community organizations

▶ Exhibit the behavior you expect in others

▶ Understand fundamental differences in individual values

▶ Match the person with the task

▶ Celebrate achievements

Evaluate your managing and leading potential by comparing your personal profile to that of the highly accomplished engineer, Arthur Morgan.

▶ Morgan's achievements included authoring new drainage laws for Minnesota, founding his own engineering firm, creating the Miami Conservancy District, rescuing Antioch College, and organizing the Tennessee Valley Authority.

▶ According to biographer Clarence J. Leuba (1971), Arthur Morgan's profile included the following 14 characteristics:
- Mentally and physically active
- Observant of a wide variety of social and physical phenomena
- Reflective and organized; given to classifying and seeking cause and effect relationships
- Avoidance, during childhood, of traditional social contacts; independent and self-reliant
- Dissatisfied with the physical and social universe; imagined improvements
- Prepared well thought out action plans
- Discriminating; identified and held to enduring values while seeking improvements to that which was ephemeral
- Open-minded but not prone to readily accept what he saw and heard
- Committed to making the world a better place
- Confident that, if he used his gifts, he could succeed

- Self-disciplined in all aspects of life including physical, mental, and emotional
- Communicative; an effective speaker and writer
- Engaged with people and able to arouse their enthusiasm and loyalty
- Systematic; carefully defined problems, explored wide range of alternatives, and then facilitated a sound decision

Do not desire to fit in.
Desire to lead.
Gwendolyn Brooks

How do you stack up? Using Arthur Morgan as a benchmark, what are your managing and leading strengths and weaknesses?

Study one or more of the following sources cited in this lesson

▶ Bennis, W.G. 1989. *Why leaders can't lead—the unconscious conspiracy continues.* San Francisco: Jossey-Bass Publishers, p. 17.

▶ Based, in part, on Coe, J.J. 1987. "Engineers as managers: some do's and don'ts." *Journal of Management in Engineering* (Oct.), pp. 281-287.

▶ Covey, S.R. 1990. *The 7 habits of highly effective people.* New York: Simon & Schuster.

▶ Kanter, R.M. 1993. Personal communication, February 22.

▶ Leuba, C.J. 1971. *A road to creativity—Arthur Morgan—engineer, educator, administrator.* North Quincy, Mass.: Christopher Publishing House.

▶ Parkin, J. 1997. "Choosing to lead." *Journal of Management in Engineering* (Jan./Feb.), pp. 62-63.

▶ Walesh, S.G. 2000. *Engineering your future: the non-technical side of professional practice in engineering and other technical fields,* 2nd ed. Reston, Va.: ASCE Press. (Chapter "Introduction," explains the leading-managing-producing model outlined in this lesson, and Chapter 15, "The Future and You," describes leadership elements.)

Refer to one or more of the following supplemental sources

▶ De Pree, M. 1989. *Leadership is an art.* New York: Dell Publishing Co.
Emphasizes the importance of identifying, freeing up, and focusing on the diverse aspirations, talents, and skills present in most organizations and groups.

▶ Gerber, R. 2002. *Leadership the Eleanor Roosevelt way: timeless strategies from the first lady of courage.* Upper Saddle River, N.J.: Prentice Hall.
Offers leading lessons drawn from Eleanor Roosevelt's life, including developing empathy, finding a mentor, taking action in crises, finding one's passion, embracing risk, and becoming an effective speaker.

▶ Hunter, J.C. 1998. *The servant: a simple story about the true essence of leadership.* Rocklin, Calif.: Prima Publishing.
Distinguishes between power, which tends to be attached to a position, and authority, which is earned by and associated with an individual. Asserts

that the latter is preferable for leading but recognizes that the former must occasionally be used. Asks if we, as aspiring managers and leaders, are getting in the way or getting obstacles out of the way. An overriding theme of this book is that to lead, we must serve the needs of others.

▶ Kotter, J.P. 1999. *John Kotter on what leaders really do.* Cambridge, Mass.: Harvard Business School Press.

Stresses the necessity of an organization having both management and leadership and offers guidance on effecting change.

> *You young lieutenants have to realize that your platoon is like a piece of spaghetti. You cannot push it. You have to get out front and pull it.*
>
> Gen. George S. Patton, Jr.

▶ Maxwell, J.C., and Z. Zigler. 1998. *The 21 irrefutable laws of leadership.* Nashville, Tenn.: Thomas Nelson Publishers.

Offers many insightful leadership "laws." Examples are influence is the true test of leadership, distinguishing between positional and real leaders, touching hearts before asking for a hand, and the power of momentum.

▶ Weingardt, R.G. 1997. "Leadership: the world is run by those who show up." *Journal of Management in Engineering* (July/Aug.), pp. 61-66.

As suggested by the title, just showing up at our own initiative, that is, when we don't have to, positions us to lead.

Subscribe to one or more of these e-newsletters

▶ "CCL's e-Newsletter," a free monthly newsletter from the Center for Creative Leadership. Provides "tools, tips, and advice to practicing managers who face the daily challenges of leading." Includes short articles and presents case studies and results of polls. Offers programs, products, and publications. To subscribe, go to http://www.ccl.org.

▶ "FMI Leadership Group e-News," a free monthly newsletter of FMI: Management Consultants to the Construction Industry. Approaches leadership development in a variety of ways, including coaching tips, survey results, and insightful quotes. To subscribe, go to http://www.fminet.com/lienews/#top.

▶ "Leadership Wired" is a free semi-monthly newsletter from John C. Maxwell. Typically included are short articles, book reviews, and quotes. To subscribe, go to http://www.injoy.com.

▶ "The Pastor's Coach" is a free monthly newsletter created by Dan Reiland, dedicated to equipping the leaders of today's church. Although obviously intended primarily for pastors and other church leaders, it contains solid, principled leadership advice applicable beyond the church setting. To subscribe, go to www.INJOY.com/PC.

Visit one or more of these websites

▶ "American Management Association" (http://www.amanet.org). The AMA, which was founded in 1923, is "dedicated to building the knowledge, skills and behaviors that will help business professionals and their organizations grow and prosper." Leads users to education and training opportunities, including seminars, conferences, forums, books, research, and self-study courses. Includes many previously presented short, application-oriented articles.

▶ "Academy of Management" (http://www.aomonline.org/). Established in 1936, the AOM is "a leading professional association for scholars dedicated to creating and disseminating knowledge about management and organizations." Includes meeting announcements, products and services, and a means for retrieval of articles previously published in the academy's journals.

▶ "Center for Creative Leadership" (http://www.ccl.org). The mission of the CCL is "to advance the understanding, practice and development of leadership for the benefit of society worldwide." Describes the center's leadership program, products, and leadership conferences and explains the function of various special groups.

A leader is a dealer in hope.

Napoleon Bonaparte

2

Roles—Then Goals

> *One man cannot do right in one department of life*
> *whilst he is occupied doing wrong in other departments.*
> *Life is one indivisible whole.*
> Mahatma Gandhi

*I*n the first part of the new year, many of us make, and often break, annual resolutions or goals. Our good intentions don't pan out. Goals deemed worthy on January 1 no longer seem to warrant special efforts. What's wrong?

One answer may be that, while our resolutions and goals are well intentioned, they lack context or perspective. Our frustrating failure frequently follows from lack of relevance of the resolutions and goals to our overall life. Our resolutions and goals may be too narrowly focused, most likely on our job, to the exclusion of our other areas of responsibility and opportunity. As a result, our goals are out of "sync" with our total being—our true range of abilities, interests, and aspirations. Perhaps our goals are solely or overly job focused because we think it's the pragmatic or responsible thing to do. After all, we "must be practical," especially as can-do technical professionals!

One way to enhance the relevance of our goals is to cast them in terms of our desired roles. Adopt a holistic approach. That is, we should first select our key roles in life, at least for the foreseeable future, and then make resolutions or establish goals that will help us fulfill those valued roles. This roles-first–goals-second idea comes from the Covey Leadership Center. Speaker and author Stephen Covey (1990) explained it this way:

> One of the major problems that arises when people work to become more effective in life is that they don't think broadly enough. They lose the sense of proportion, the balance, the natural ecology necessary to effective living. They may get consumed by work ...

An example of a non-work role that is likely to be shared by many technical professionals is that of parent. Other common non-work roles are daughter, son, wife, husband, grandparent, neighbor, athlete, friend, member of a religious group, and community leader. As used in the preceding sentence, "common" means a role likely to be held by many individuals, as opposed to a rare role. "Common" does not mean unimportant.

Examples of work-related roles likely to be held by engineers and other technical professionals are designer, administrator, marketer, project manager, partner, mentor, committee member, and officer in a professional, business, or service organization. Frankly, today's typical technical professional probably has more work-related roles than his or her counterpart did a decade or so ago. Such added expectations are one manifestation of the changing world of work. All the more reason for each of us to perform a role check.

Although many of us share some roles, our interpretations of success in any given role will vary widely. Consider, for example, the non-work role of a community member. Some of us probably think that we successfully fulfill our community responsibilities by quietly going about our business. Others would say that success in the community role requires proactive involvement in our immediate neighborhood or perhaps an even higher profile leadership role in the community at large. All are legitimate, especially if they are done by design.

> *Purpose is what gives life meaning.*
> Charles Henry Parkhurst

Clearly, we can establish goals without first defining roles. The danger is that we will inadvertently omit or diminish important segments of our being. We risk incurring deep regrets that cannot be remedied. In contrast, the suggested "roles—then goals" process has the advantage of causing each of us to strive for balance in our life. Thomas L. Brown (1986), management writer and speaker, said it this way, ". . . you cannot balance your personal and professional life unless there is substantial weight on both ends."

Suggestions for Applying Ideas

Experiment with the roles-first–goals-second idea advocated in this lesson by at least starting the process. Create and fill out a table like the following

| Roles | Goals | Action Items | | Status |
		What?	By When?	
Active citizen	Serve on community committee	Identify existing/ proposed committees	10/31/2003	Searching
		Prioritize	11/30/2003	
		Volunteer	12/31/2003	

▶ Refer to the essay portion of this lesson for a list of possible entries in the Roles column.

▶ Just the process of preparing the preceding table can be revealing, especially if, at the outset, you have very few rows.

▶ If you like what you see, such as the possibility of more balance in your life, experiment with the table for at least a month. Update the table weekly and use it to schedule some of your weekly activities. See if your behavior changes and if improved balance is achieved.

Read the following related lessons

▶ Lesson 3, "Smart Goals"

▶ Lesson 13, "Afraid of Dying, or Not Having Lived?"

▶ Lesson 49, "Giving to Our Profession and Our Community"

Study one or more of the following sources cited in this lesson

▶ Brown, T.L. 1986. "Time to diversify your life portfolio?" *Industry Week* (Nov. 10), p. 13.

▶ Covey, S.R. 1989. *The 7 habits of highly effective people.* New York: Simon & Schuster.

Refer to one or more of the following supplemental sources

> *I don't intend to be commonplace. I intend to make a great person of myself... great in having fulfilled my possibilities: great in having seen which of my possibilities are greatest.*
>
> Arthur Morgan

▶ Leuba, C.J. 1971. *A road to creativity—Arthur Morgan—engineer, educator, administrator.* North Quincy, Mass.: Christopher Publishing House.

Explains how Arthur Morgan, born of modest means, but with a stimulating home environment, creatively sought out and succeeded in a wide variety of professional roles. Describes his accomplishments as a water control engineer, founder of an engineering firm, creator of the Miami Conservancy District, rescuer of Antioch College, and organizer of the Tennessee Valley Authority.

▶ Lorsch, J.W., and T.J. Tierney. 2002. "Build a life, not a resume." *Consulting to Management* (Sep.), pp. 44-52.

Offers advice supportive of the roles theme of this lesson. Advocates personal alignment, that is, making decisions that mutually reinforce our "capabilities, goals, needs, and values."

Subscribe to one or more of these e-newsletters

▶ "Making a Life, Making a Living" is a monthly e-newsletter produced by Mark S. Albion. Uses questions, humor, quotes, and short essays to thoughtfully address a wide variety of personal decision topics. To subscribe, go to http://www.makingalife.com

> *Philosophy is*
> *the art of living.*
> Plutarch

▶ "Winner's Circle Daily Email" is a free e-newsletter provided by the Pacific Institute. Example topics are life strategy, loneliness, goals, coping, and parenting. To subscribe, go to http://mailman.wolfe.net/mailman/listinfo/wcn.

Visit this website

▶ "Making a Life, Making a Living" (http://www.makingalife.com/) is maintained by Mark S. Albion. As suggested by the title, meaningful living is stressed. This website markets products and includes a free quote search feature.

Building an impressive resume is a lot easier than building a fulfilling life because life is a lot more complicated.

Jay W. Lorsch and Thomas J. Tierney

Smart Goals

Those who do not have goals are doomed to work for those who do.
Brian Tracy

*L*esson 2, "Roles—Then Goals," argues that we should decide on our high-priority roles (e.g., parent, professional, daughter, friend) before we establish our goals. Why? The danger of establishing goals before defining roles is that we will inadvertently omit or diminish important segments of our being. We risk incurring deep regrets that cannot be remedied. Clear goals, consistent with our selected roles, are crucial to charting and navigating the tumultuous seas of our professional, community, and personal life. As stated by publisher Malcolm Forbes: "If you don't know what you want to do, it's harder to do it."

As a guide to formulating annual or other goals, consider using the acronym **SMART**.

S means be *S*pecific. A vague goal, such as "become a better member of the community" isn't very helpful. Instead, try to be more specific: "become an active member of a Chamber of Commerce committee."

M stands for *M*easurable. To the extent feasible, each goal should be cast in quantitative terms. An example is "complete 90% of my projects under budget." That which is measurable is more likely to get done.

A refers to *A*chievable. While we should be stretched by goals, we must be able to accomplish each one assuming a sustained, good faith effort. A major goal, such as publishing a book, could be broken into ambitious, achievable sub-goals, one of which might be "sign a contract with a publisher."

R denotes *R*elevant. Each goal must be relevant to your chosen roles and other constraints. Establishing a goal of starting a service or product line in your company that is at odds with the organization's strategic and business plans fails the relevance test.

T represents *T*ime-framed. Establish a schedule or set milestones for achieving a goal or its components.

We are most likely to fulfill our chosen roles if we are guided by well-formulated goals. Are your goals SMART, that is, specific, measurable, achievable, relevant, and time-framed?

Implicit in the preceding advice is that our goals should be written. The suggested SMART technique can only be fully implemented in writing. Conceptualize, refine, and write out monthly, annual, and multi-year goals for personal, family, financial, community, and professional areas and affairs.

Having written our goals, we should frequently look at them. Each of us has control over how we invest some of our time and energy. Important, urgent matters tend to dominate our lives. Working toward goals falls in the important but not urgent category. As such, our goals can wither from lack of attention while we focus our care on urgent demands. We must invest some of our spare moments in reviewing our goals and planning action items to achieve them. The leader in us schedules time to review and refine personal and group goals. Ralph Waldo Emerson, schoolmaster, minister, lecturer, and writer, said:

> Guard your spare moments. They are like uncut diamonds. Discard them and their value will never be known. Improve them and they will become the brightest gems of a useful life.

A final thought. Selectively share goals with trusted individuals who can help you achieve your objectives. For example, if one of your goals is to work on a particular type of project, discuss the details with your supervisor and others in influential positions. Perhaps you desire to serve on a committee or task force in your community. Then discuss your interest with appropriate elected or appointed community officials. You are surrounded by people who, if they know you and your goals, will help you achieve them.

It's never too late to be what you might have been.
George Eliot

Suggestions for Applying Ideas

Write personal SMART goals in each of the following areas for 1 year and 10 years from now

I've always wanted to be somebody, now I realize that I should have been more specific.
Lily Tomlin

► Annual salary and other compensation.

► Position, such as project engineer, project manager, owner, and independent consultant.

► Functions, such as design, marketing, manufacturing, and general management.

► Other, such as travel internationally, start a business, serve in elected office, and publish paper and/or book.

Identify, for each of the preceding SMART goals, one specific thing you will do this month, for the 1-year goals, and this year, for the 10-year goals

▶ By setting a goal you are, in effect, "planning a trip." How are you going to get to your destination?

▶ Do you have the necessary knowledge, skills, and attitudes and, if not, a means of obtaining them?

▶ Or are you going to let chance rule, perhaps using the rationale that everything will come to you if you "just work hard" and "keep your nose to the grindstone"?

Manage your time wisely, so that you devote sufficient energy to achieving the established SMART goals. Consider these time management ABCs (Walesh, 2000)

▶ Articulate SMART goals—presumably you have already done this.

▶ Plan each day in writing.

> *A winner is someone who recognizes his God-given talents, works his tail off to develop them into skills and uses these skills to accomplish his goals.*
> *Larry Bird*

▶ Act immediately and constructively. For example, if a communication comes across your desk, immediately act on it, file it, or discard it.

▶ Bring at least one potential solution whenever you take a problem to a supervisor, colleague, or client. Expect your supervisees and others to do the same.

▶ Identify your best time of the day and use it for your most demanding tasks.

▶ Maintain a clean desk and work area to minimize visual distractions.

▶ Create an efficient workspace with frequently used items in easy reach.

▶ Distinguish between efficiency (doings things right) and effectiveness (doing the right things). Both are necessary, but the latter is more important. Goals help us identify the right things.

▶ Create a carefully organized set of paper and/or electronic professional files in which you place potentially useful material for possible future retrieval and use.

▶ Keep materials for ongoing small projects together. Clear plastic kitchen storage bags provide one way to do this.

▶ Meet only when necessary, and, to the extent you have influence, insist that meetings be carefully planned and facilitated.

▶ Recognize the 20/80 rule of thumb: 20% of the input to a process produces 80% of the results. Search for and concentrate on input items in the 20%.

▶ Break large projects into small, manageable (by you or others) parts.

▶ Use discretionary time wisely, recognizing that time at work may be viewed as falling into one of these three categories: boss-imposed time, system-imposed time, and self-imposed time. Self-imposed time is discretionary time and is most likely to offer opportunities to advance your goals by attending to important but not urgent demands.

> *Devoting a little*
> *of yourself*
> *to everything means*
> *committing a great*
> *deal of yourself*
> *to nothing.*
> Michael Le Boeuf

▶ Group similar activities such as returning telephone calls or e-mails or working on various parts of a report. Grouping tends to be more efficient.

▶ Avoid "telephone tag" by using techniques such as leaving a specific voice mail message, leaving an intriguing message, suggesting a specific telephone meeting time, and using other forms of communication.

▶ Delegate appropriate parts of your tasks, along with necessary authority, to other capable individuals.

▶ Keep "door" closed but access open. You are less likely to be interrupted if your door is closed or the entry to your workspace is arranged such that potential visitors interrupt only when necessary.

▶ Write it down. Time invested in documentation efforts, such as taking notes at a meeting, writing a memorandum to file, or sending an e-mail to meeting participants, usually improves understanding of decisions and responsibilities and saves time in the long run.

▶ Write responses on original documents. Answer selected hardcopy memoranda and letters by writing a response directly on the document and returning it to the sender, perhaps with copies to others and the file.

> *We're lost, but we're*
> *making good time.*
> Yogi Berra

▶ Use travel and waiting time productively. Carry small reading or other projects with you, along with appropriate tools.

▶ Use word processing, as opposed to writing longhand, but recognize that a handwritten note or card, because of the rarity, can be a very meaningful means of communication.

▶ Meet with yourself, that is, isolate blocks of time during your workday to do tasks requiring higher concentration.

▶ Log your time. Keep a record, at 15- to 30-minute intervals, of how you use your time for several days to a week. Identify and expand productive uses.

▶ Adopt a holistic philosophy by striving for balance among the intellectual, physical, emotional, and spiritual dimensions of your life.

▶ Guarantee small successes. Plan each day so it includes one activity that is both enjoyable and likely to be accomplished.

Be prepared to persevere to achieve your more ambitious goals. Consider these examples of perseverance

▶ Chester Carlson developed the quick electrostatic photography process in the 1940s. This technology was intended to replace the contemporary, cumbersome, copying paradigm, which used film, developer, and a darkroom. Incredibly, 43 companies rejected his idea! They passed on the opportunity to develop what is now called xerography and is the basis for the omnipresent copy machines (Barker, 1989).

▶ Theodor Geisel, more popularly known as Dr. Seuss, is considered a premiere author of children's books. He was a pioneer in linking drawings to text, an approach that appeared in his first book. His first book was rejected by 29 publishers before being accepted (http://www.carolhurst.com).

▶ Joseph B. Strauss, engineer and poet, envisioned a bridge across San Francisco's Golden Gate. In the face of widespread skepticism, including that of his peers, he led the two-decade planning, design, and construction of the Golden Gate Bridge. Strauss, who died approximately one year after the May 1937 opening of the bridge, is honored with a statue erected at the south end of the span, dedicated to "The Man Who Built the Bridge." The intensity of that effort is suggested by these lines from one of Strauss' poems (McGloin, 2003):

People are always blaming their circumstances for what they are. I don't believe in circumstances. The people who get on in this world are the people who get up and look for the circumstances they want, and, if they can't find them, make them.
George Bernard Shaw

> Launched midst a thousand hopes and fears,
> dammed by a thousand hostile sneers.
> Yet ne'er its course was stayed.
> But ask of those who met the foe,
> who stood alone when faith was low,
> ask them the price they paid.

Unfortunately, Strauss' admirable perseverance was partly offset by his reluctance to acknowledge the contributions of others. For example, Charles A. Ellis, who actually led the design effort, was discharged by an apparently highly egotistical Strauss before construction was completed (McGloin, 2003; Fredrich, 1989).

Read the following related lessons

▶ Lesson 2, "Roles—Then Goals"

▶ Lesson 9, "Go Out on a Limb"

▶ Lesson 13, "Afraid of Dying, or Not Having Lived?"

▶ Lesson 32, "The Power of Our Subconscious"

▶ Lesson 33, "Delegation: Why Put Off Until Tomorrow What Someone Else Can Do Today?"

▶ Lesson 52, "Looking Ahead: Can You Spare a Paradigm?"

Study one or more of the following sources cited in this lesson

▶ Barker, J.A. 1989. *Discovering the future: the business of paradigms*. St. Paul, Minn.: ILI Press.

▶ Fredrich, A.J. (ed.). 1989. Strauss gave me some pencils. In *Sons of Martha: civil engineering readings in modern literature*. New York: ASCE.

▶ Hurst, Carol http://carolhurst.com. A collection of reviews of children's books.

▶ McGloin, J.B. 2003. "Symphonies in steel: Bay Bridge and the Golden Gate." Museum of San Francisco website: http://www.sfmuseum.org/hist9/mcgloin.html.

▶ Walesh, S.G. 2000. *Engineering your future: the non-technical side of professional practice in engineering and other technical fields*, 2nd Ed. Reston, Va.: ASCE Press. (Chapter 2: "Management of Self.")

Refer to one or more of the following supplemental sources

▶ Hensey, M. 1999. *Personal success strategies: developing your potential*. Chapter 14, "Setting work, career and personal goals." Reston, Va.: ASCE Press.

Introduces the idea of requesting feedback from bosses, colleagues, clients and others as part of the process of defining goals. Example questions might be: What personal liability should I work on? What strength should I enhance? What skills should I sharpen? What knowledge would help me?

▶ Hill, N. 1960. *Think and grow rich*. New York: Fawcett Crest.

Notes the power of "creating, in your own mind, a burning desire" for each goal and emphasizes the importance of written goals and plans to achieve them.

▶ Urban, H. 2003. *Life's greatest lesson: 20 things that matter*. New York: Simon & Schuster.

Defines success as "the progressive accomplishment of worthy goals." Asks the question, "we draw up plans for buildings, businesses, meetings, weddings, sports, parties, vacations, retirement, etc. But do we draw up plans for our lives?"

Subscribe to one or more of these e-newsletters:

▶ "Personal Achievement Quote of the Day" is offered free by the company Top Achievement. Uplifting thoughts from varied accomplished individuals may help you implement your goal-related action plans. To subscribe, go to http://www.topachievement.com/quote.html.

▶ "Winner's Circle Daily Email" is provided free by the Pacific Institute. The institute and the e-newsletter teach "people how to manage change, set and achieve goals, lead more effectively and think in ways that create success." To subscribe, go to http://mailman.wolfe.net/mailman/listinfo/wcn.

Visit this website

▶ "Top Achievement" (http://www.topachievement.com/) is the website of the company Top Achievement. Provides many free goal-related articles and access to websites and products.

You must have long-term goals to keep you
from being frustrated by short-term failures.

Charles Noble

4

Too Much of a Good Thing

*The trouble with experience is that by the time you have it
you are too old to take advantage of it.*

Jimmy Connors

A downturn in the stock market as indicated, for example, by a sharp drop in the Dow Jones Industrial Average, is disconcerting for many of us. Suddenly retirement accounts, mutual funds, and other investments plummet in value. Our net worth drops sharply. There is the gnawing fear of the negative long-term effect on the material well-being of our spouse and dependents. We vow to be even more watchful of our investments. After all, prudence requires careful management of personal financial assets.

What about the status of and attention to personal professional assets? Although the value of our professional assets defies quantification, it is nevertheless very real. The quality of our experience is a major part of those assets. Experience gives us the ability and confidence to take on new challenges, including more managing and leading. Thoughtfully applied, experience also helps us learn from and not repeat mistakes.

*Many people end up
in the wrong place
only because
they stayed in the
right place too long.*

John C. Maxwell

While experience is valuable, too much of one kind of experience can hamper individual growth. Accordingly, each of us should appraise our professional assets at least once a year, in part to assess the quality and freshness of our experience.

This evaluation of professional assets might be in the form of a resume update exercise. What new areas of technology have been mastered? What new management and leadership techniques were used? What new concepts, ideas, or principles were studied? How has our attitude improved? What new skills were acquired? What new challenges and responsibilities were accepted? What new opportunities were seized and new risks taken? What knowledge was shared with co-workers and others? What new contributions were made? In what ways have we been "good and faithful servants" with our talents?

One thorn of experience is worth a whole wilderness of warning.
James Russell Lowell

As we review several annual accountings of our professional experience, will we find several years that each were filled with new experiences? Or will we find one year of experience repeated several times? If the latter is true, we may be in the midst of a devastating professional assets "stock market crash."

The value of professional assets, if approximated by the present worth of one's future income, is likely to exceed the value of a younger person's net worth, except for those fortunate few who are independently wealthy. Like personal financial assets, personal professional assets can appreciate, plateau, or decline.

We should resist the temptation to settle into the comfort of routine, rationalizing it in the name of gaining more experience. Author and lecturer Og Mandino, in his book *The Greatest Salesman in the World* (1968), offers this warning about excessive experience:

> I will commence my journey unencumbered with either the weight of unnecessary knowledge or the handicap of meaningless experience. . . . In truth, experience teaches thoroughly yet her course of instruction devours men's years so the value of her lessons diminishes with the time necessary to acquire her special wisdom.

Claiming "I need more experience before I . . ." often reflects a sincere desire to be properly prepared before taking on a new task, function, or other challenge. In some cases, the claim that more experience is needed reflects deep-seated fear of advancement; it becomes a rationalization for staying put. Clearly, as suggested by Mandino's advice, we need to strike a balance between too little and too much experience. Using a sailing metaphor, author Richard Bode (1993) gives this advice for achieving balance:

> The frantic individual tacks too soon, jumping from job to job. . . . The obtuse individual remains on the same tack too long, investing too much time, talent and energy in a course that takes him far from his avowed objectives. But the seasoned sailor stays on the same tack as long as it appears advantageous, and then . . . deftly changes directions . . . the confirmed sailor goes on tacking forever.

Each of us has opportunities to deftly change direction so that we gain new and valuable professional experience. Examples of ways to acquire asset-building managing and leading experience include asking for new job assignments, requesting a transfer to another part of our organization, seeking a new employer, returning to school, establishing our own business, and becoming active in a professional or business organization. In the final analysis, each of us has his or her hand on the tiller.

Suggestions for Applying Ideas

Apply the Shewart Cycle (Hensey, 1999) as a means of maximizing what you learn from experiences

- ▶ The Shewart Cycle is the plan-do-study-act process of continuous improvement:
  - Plan a step or steps toward a goal.
  - Do as per the plan.
  - Study the results.
  - Act to improve the step or steps and repeat the cycle.

- ▶ The value of applying the Shewart Cycle to experience is that it emphasizes the need to examine our experiences. Experience in and of itself is of no value. We must examine our experiences to avoid missing opportunities to learn. Rather than repeatedly having the same experience, or kind of experience, the Shewart Cycle helps us benefit from our experience and, therefore, move on to different and higher level experiences.

> *Prove to your own satisfaction,*
> *that every adversity, failure, defeat, sorrow and unpleasant circumstance, whether of your own making or otherwise, carries with it the seed of an equivalent benefit which may be transmitted into a blessing of great proportions.*
> Napoleon Hill

- ▶ Consider a professional experience you are about to have. Examples might be designing a product, interviewing for a new position, giving a talk, calling on a potential client, and attending a meeting. Consciously apply the Shewart Cycle as follows:
  - Plan the experience, at least the aspects you control or influence.
  - Do (that is, carry out) the plan.
  - Study, and learn from, the results.
  - Act: decide what you would do the same and differently the next time, and then do those things that way when the opportunity arises.

Look to your immediate surroundings for new profession-related experience opportunities

- ▶ Someone in a professional rut may prematurely decide that new experiences require a new employer. While the grass may be greener on the other side of the fence, green patches may already lie on your side. Consider these possibilities within your current employment situation:
  - Share some or all of your goals with a supervisor, and ask for a challenging assignment consistent with one or more of your goals.
  - Request a transfer to another department, office, or other organizational unit. Even if financial compensation remains the same, the range and variety of experiences is likely to grow and eventually lead to more opportunities and increased compensation.

- Offer to accompany others who are attending meetings, visiting project sites, and carrying out other functions. Vicariously experience their decision-making, presentations, discussions, and other actions.

In the business world, everyone is paid in two coins: cash and experience. Take the experience first; the cash will come later.
Harold Geneen

Consider the similarity between learning from experience and digesting food (Chopra, 2001)

▶ Our bodies extract nourishment from food, even when we don't eat well, and eliminate what is not needed.

▶ Similarly, we can learn to extract knowledge from our positive and negative experiences and let go of the residue, that is, whatever does not serve us.

Read the following related lessons

▶ Lesson 6, "Defining Moments"

▶ Lesson 10, "Keeping Our Personal Financial Score"

▶ Lesson 15, "So What Do You Know about Bluebirds?"

Study one or more of the following sources cited in this lesson

▶ Bode, R. 1993. *First you have to row a little boat.* New York: Warner Books.

▶ Chopra, D., and D. Simon. 2002. *Grow younger, live longer: 10 steps to reverse aging.* New York: Three Rivers Press.

▶ Hensey, M. 1999. Chapter 6, "Learning How to Learn." In *Personal success strategies: developing your potential.* Reston, Va.: ASCE Press.

▶ Mandino, O. 1968. *The greatest salesman in the world.* New York: Bantam Books.

Refer to the following supplemental source

▶ Hill, N. 1960. *Think and grow rich.* New York: Fawcett Crest.
Reports on a 25-year study of the experiences of highly accomplished individuals. Concludes that the common element is the power of visualization—looking way beyond one's current situation—linked with the subconscious mind.

From time to time, everyone benefits from being "re-potted," from applying their talents to new challenges. Re-potting and self-renewal go hand in hand, whether the pot is a new position, a new firm or an entirely new career.
Jay W. Lorsch and Thomas J. Tierney

5

DWYSYWD

It is better that you should not vow than that you should vow and not pay.

Eccles. 5:5, Bible, RSV

Promises punctuate our professional and private lives. They are mostly small, with big ones intermingled, but nevertheless all are promises—or at least seem to be. We could easily hear a dozen or more promises on any given day. However, informed by experience, we gradually learn that there are "promises" and there are promises. The former are just words, while the latter are commitments to action; they are as good as done.

One way we risk engendering mistrust, tarnishing our reputations, and frustrating our aspirations is by failing to keep small promises. For example, you meet someone at a local meeting of a professional society, you exchange business cards, and promise to send her a copy of your firm's brochure. But you forget. Or you run into an acquaintance that you have not seen for some time, talk briefly, and agree that you and he should get together for lunch. You offer to make arrangements. But you forget. Other examples of promises often not kept are listed below:

- *I'll send* you a copy of the material.
- Send me the draft report; *I'll give* you my comments.
- *I'll draft* the minutes; you will have them within three days.
- Don't worry about the sizes of the structural columns; *I'll check* the calculations.
- *I'll contact* the Department of Natural Resources and get back to you.

Breaking promises like those cited here might be considered insignificant oversights. But are they harmless, especially if part of a pattern?

We might argue that breaking small promises is not a significant issue. Small promises are in the gray area, you say. Perhaps. After all, there are no explicit "small promises" canons, rules, principles, or tenets in the various ethics codes governing professional practice in engineering. The immediate consequence of any unfilled small promise, such as not reading a brochure or missing lunch, is usually small. Furthermore, the technical professions are built on a rational foundation of science and technology. Small, interpersonal failings would seem to pale in importance to the hard science and technology that the technical professions wield.

However, getting opportunities to effectively apply that technology in the form of products and services that others are willing to pay for or invest in requires developing a web of mutually trustful relationships with co-workers, business partners, clients, elected officials, regulators, and others. Clearly, major transgressions are likely to permanently shatter trust or prevent it from occurring. However, and this is the point of this lesson, cumulative small lapses can have the same effect. They gradually exclude us from the web of those who have faith and confidence in each other, and we miss managing and leading opportunities.

Those who routinely don't do what they say they will do are sometimes creative and well-intentioned. However, these positive qualities do not lead, as they could, to positive outcomes because the creative, well-intentioned individuals are poor managers of their personal time and affairs. They lack the discipline to write down, on paper or electronically, their promises and periodically review and act on their "to do" list. As a result, those they interact with miss out on the fruits of creative thoughts.

If someone throws you a ball, you don't have to catch it.
Richard Carlson

Others who don't do what they said they would do make promises because they become entrapped, in knee-jerk fashion, by a desire to please. They promise because it seems to be the thing to do. They haven't learned how to say no or to say nothing. Their promises are shallow and lack conviction.

Regardless of the reasons for frequently not following through on promises, which can be many and varied, the fate of those who don't is common: loss of credibility. Family, friends, colleagues, clients, potential clients, and others gradually recognize the wide gulf between what the person "promises" and what he or she delivers. We are increasingly reluctant to rely on certain individuals and may "write off" others after concluding that their promises, regardless of motivation, are just vibrations in the air. To the extent feasible, we avoid entanglement in their webs.

Talk is cheap to some of the people we encounter as evidenced by the lavish way they dispense words. Enthusiasm abounds, ideas flow, proposals are presented, and promises are made. However, experience with these individuals gradually reveals that promises, big and small, are not kept. While we may continue contact with the "talkers," they are not likely to be part of our trusted circle of colleagues. We begin to question their management competency and leadership potential.

Successful manufacture or building of a product, structure, facility, or system requires a careful synthesis of components. The size of any component doesn't necessarily determine its importance. Recall the innocuous O-rings that precipitated the Challenger disaster and the deceptively simple, small hanger rod and box beam connections that caused the fatal Kansas City Hyatt walkway collapse. Similarly, successful development of mutually trustful relationships demands attention to both the big picture and the little details. Trust is built piece by piece; some pieces are large but many are small. By keeping small promises, we build big relationships.

> *No virtue is more universally accepted as a test of good character than trustworthiness.*
>
> Harry Emerson Fosdick

We engineers should do whatever is necessary to earn and retain client, customer, citizen, and other stakeholder trust. Keeping promises, including little ones, is essential to trustful relationships. My advice: DWYSYWD. Do what you said you would do.

Suggestions for Applying Ideas

Follow up immediately and, if possible, in writing, on each promise

> *Do what you said you were going to do, when you said you were going to do it, and how you said you were going to do it.*
>
> Byrd Baggett

▶ If you met with someone and promised to send him a document, draft a transmittal letter, memorandum, or e-mail, and/or obtain the documents as soon as you get back to your office. The point: do something to start the process of following through on your promise.

▶ If you promised to arrange a meeting with someone, place a reminder in your time management system to contact her. Make the entry in her presence; it suggests your commitment.

Consider this advice on promise-making from Robert Townsend (1970), the blunt-speaking former CEO of the Avis car rental company

▶ If asked when you can deliver something, ask for time to think.

▶ Build in a margin of safety.

▶ Name a date.

▶ Then deliver it earlier than you promised.

Read the following related lesson

▶ Lesson 3, "Smart Goals" (the time management portion)

Study the following source cited in this lesson

▶ Townsend, R. 1970. *Up the organization: how to stop the corporation from stifling people and strangling profits.* New York: Alfred A. Knopf.

You cannot live on other people's promises,
but if you promise others enough, you can live on your own.
Mark Caine

6

Defining Moments

Experience is the hardest kind of teacher.
It gives you the test first and the lesson afterward.
Anonymous

Several decades of work and other experiences prompt the question, "Why do we think and act the way we do?" Each of us possesses a special set of strong principles, traits, and skills that influence our thinking and guide our actions. What is their origin?

Some of what we are professionally and otherwise is traceable to what can be called defining moments. Leadership and change authors W.G. Bennis and R.J. Thomas (2002) also use the term defining moments, as well as turning points, and introduce the crucible as the metaphor for the defining moment and its surrounding circumstances. Defining moments tend to be surprising, riveting, emotional, and sometimes seemingly catastrophic at the personal level. In short, defining moments are memorable. But, more importantly, depending how we react, they can have profound influence on how we subsequently think and what we subsequently do.

An optimist sees an opportunity in every calamity; a pessimist sees a calamity in every opportunity.
Winston Churchill

To illustrate the point, let me share five personal defining moments, each of which ultimately shaped me in a positive fashion.

▶ While an engineering student, I was diligently working on an exercise in a freshman engineering graphics class. We students and our instructor were suddenly interrupted by the dean. He burst into the classroom, requested our attention, and said something like: "The four forms of communication are speaking, writing, mathematics, and graphics. You should strive to be competent in all of them. Thank you." From that defining moment on, I realized the importance of communication and committed to developing the skills.

▶ About 40 years ago, I was in a surveying laboratory. The instructor was trying to show us how to roll up (throw) a steel chain, a type of tape measure used by surveyors. He was right-handed and I was left-handed. When my turn came, I did it incorrectly and he belittled me in front of the class. Drawing on that defining moment, I strove in my teaching work to always expect much of students but to never belittle them in front of their peers, or in private.

▶ A goal, at one point in my career, was to become head of one of our firm's offices. The company announced that the current manager would soon move into a new position. The position I desired would be vacant. I thought I had done all the right things, including working hard, holding a variety of positions, and starting to plan what I would do as office manager. On the day the new manager was announced and it wasn't me, I quickly went through shock, disappointment, anger, and self-doubt. But that defining moment caused me to step back, obtain a broader perspective, and launch into a new, satisfying career. Another lesson: share goals with decision makers.

Treat pain and rage as visitors.
Ben Hecht

▶ I once shared an idea, having major implications if implemented, with my boss. Because it was an unusual idea and had major ramifications for my area of responsibility, I asked him to think about the idea while holding it in confidence until we could meet again. At a large, high-level meeting a week or so later, my idea, minus a critical element, was presented as the boss' proposal and adopted. I felt betrayed and knew that I could no longer trust the person I reported to. Largely as a result of that defining moment, I resigned my position and moved into a successful and satisfying vocation I had only dreamed of doing.

▶ About 30 years ago, my wife and I were traveling by air. I couldn't help but overhear the conversation of an apparently married couple across the aisle, who, at that time, appeared to be rather elderly. (Let's call them Ralph and Muriel.) At the end of the conversation, Ralph turned to Muriel and said something like, "We had all these hopes—and now it's too late." The sadness and finality of Ralph's comment was, for me and my wife, a defining moment. We subsequently resisted rationalizations to defer, in knee-jerk fashion, ambitious ideas and aspirations.

Each of us will occasionally be confronted with surprising, emotionally jolting, and apparently catastrophic experiences. They can occur in our professional, community, and family affairs. They may initially knock us down and set us back.

Every adversity carries with it the seed of an equal or greater benefit.
W. Clement Stone

While we should fully experience and selectively share our emotions at these very troublesome times, experience suggests that what appears to be a problem may, on reflection, define a previously unseen opportunity. Prudence suggests creatively and courageously looking at these instances as potential, positive defining moments. A valuable life-long lesson may emerge. We may strike out in a new, previously unforeseen direction.

Suggestions for Applying Ideas

Revisit defining moments in your life (Bennis and Thomas, 2002)

▶ Draw a horizontal line that begins with your birth and extends to the present.

▶ Note, along the line, your defining moments and when they occurred.

> *When one door closes another door opens; but we often look so long and so regretfully upon the closed door that we do not see the ones which open for us.*
>
> Alexander Graham Bell

▶ Recall how each defining moment appeared to you then (e.g., positive or negative, uplifting or devastating, equitable or unfair) and contrast that with how it appears now, in retrospect.

▶ For each defining moment, note the good that came out of it and consider what you would do differently if you could experience that defining moment again.

The preceding should help you respond to possible future defining moments and perhaps help you counsel others who are experiencing their defining moments.

Read the following related lessons

▶ Lesson 7, "Courage: Real and Counterfeit"

▶ Lesson 13, "Afraid of Dying, or Not Having Lived?"

Study the following source cited in this lesson

▶ Adapted from Bennis, W.G., and R.J. Thomas. 2002. *Geeks and geezers: how era, values, and defining moments shape leaders.* Boston, Mass.: Harvard Business School Press.

Refer to the following supplemental source

▶ Wiersbe, W.W. 1994. *God isn't in a hurry.* Grand Rapids, Mich.: Baker Books. Chapter 14, "Wanted Hinge People," uses the hinge as a metaphor for a defining moment or a turning point. The author and pastor says, "Just as a large door can swing on a small hinge, so a great life can be guided by a perhaps seemingly insignificant event . . . or a little known person."

> *Every good thing that has happened to me has happened as a direct result of something bad.*
>
> Harry Caray

Courage:
Real and Counterfeit

*It is curious that physical courage should be so common
in the world and moral courage so rare.*
Mark Twain

Leadership and significant achievement require courage—courage to set high personal and group goals, to keep the faith in the face of major set-backs, to hold people accountable for carrying out their responsibilities and keeping their promises, to confront individuals exhibiting unacceptable behavior, to walk away from a project or client on ethical grounds, to aim high and risk apparent great failure, to apologize and ask for a second chance, and to persist when all others have given up. But, what constitutes courage and courageous people? The aspiring leader that lies within many of us would like to know.

The Greek philosopher Aristotle (1987) offers a thoughtful and demanding perspective on courage. He defines courage as a precarious, difficult-to-describe balance between causes, motives, means, timing, and confidence:

> The man, then, who faces and who fears the right things and from the right motive, in the right way and at the right time, and who feels confidence under the corresponding conditions, is brave; for the brave man feels and acts according to the merits of the case and in whatever way the rule directs.

Aristotle goes on to say that courage is a mean between cowardice and rashness, confidence and fear. In summary, he defines courage as a fully informed, carefully considered willingness to die for a noble cause. Aristotle refutes the notion that courage is reactive or instinctive. We might be tempted to say that Aristotle was not totally serious about his definitions of courage and courageous people, at least with respect to the "willingness to die" aspect.

After all, he must have intended death as a metaphor for a willingness to incur great loss. This interpretation is probably acceptable.

Aristotle outlines in systematic and exhaustive fashion five kinds of false courage. These might be referred to as lesser degrees of courage. They encompass much of what passes for courage in our society and help, by elimination, to define bona fide courage.

▶ The first type of courage is *coercion courage*, or what Aristotle refers to as "the courage of the citizen-soldier." The possessor faces significant risks, but he or she has no choice. Leaders simply have to do many things, some of which are quite unpleasant and risky. Aristotle's coercion courage concept cautions the leader in us to maintain perspective and not to view these as courageous acts worthy of praise. These acts are part of the job; they come with the territory.

▶ What might be called high information or *calculated courage* is the second type. Aristotle uses the example of the professional soldier who seems brave in battle, but in fact entered the fray with far superior information and other resources that virtually guaranteed victory. According to World War II air ace Eddie Rickenbacker, "Courage is doing what you're afraid to do. There can be no courage unless you're scared." The leader in us may be tempted to feign courage because we have exclusive access to vital information.

> *It is easy to be brave from a safe distance.*
> Aesop

▶ The third type of courage is *passion courage*. These reactionary acts conflict with the choice and motive elements clearly evident in Aristotle's model of courage. While the emotional outburst or sharp retort is often viewed as courage—as in "you sure told him/her"—these acts are often done without thought. Although passionate reactions seem to immediately please some onlookers, calm and reason in difficult circumstances may require more courage and lead to greater long-term benefits for all antagonists.

▶ Sanguine, to use Aristotle's word, or what might be called *overly optimistic courage*, is the fourth type of counterfeit courage. A string of business or other successes can lead to unrealistic optimism or even complacency, which may be viewed as courage. The U.S. global dominance in economic and military affairs during the four-decade post-World War II period is an illustration of Aristotle's sanguine courage. The modern leader must be alert and view expectations of continued success with suspicion. An earlier atmosphere of courage that enables an organization to achieve high levels of performance may gradually and unnoticeably give way to complacency.

▶ Aristotle's fifth type of false courage is the *ignorance variety*. As he bluntly says, "People who are ignorant of the danger also appear brave." As we become an increasingly information-rich world, the leader in us must devote appropriate resources to continuously sifting through new knowledge to identify and assess opportunities and threats. At any given time, our courage may, in fact, be based on a lack of data and information describing the dire circumstances we are facing.

Far better it is to dare mighty things, to win glorious triumphs, even though checkered by failure, than to take rank with those poor spirits who neither enjoy much nor suffer much, because they live in the gray twilight that knows not victory nor defeat.

Theodore Roosevelt

Informed by Aristotle's ideas, the aspiring leader in each of us is more likely to recognize our and others' bravado. There will always be some pretense of bravery, particularly by people in high and prestigious positions. Recognizing this, the leader should place a premium on his or her acts and the acts of others that, in the face of risk and calamity, are carefully considered and indicate a willingness to sacrifice for the corporate or community cause. Courageous acts don't have to be extreme acts. When leaders take extreme positions, they may be less successful in defending a principle, advancing a cause, or achieving a worthy goal than when they assume courageous but somewhat more moderate postures. The leader in each of us recognizes various types of false courage and seeks instead a courage that balances causes, motives, means, timing, and confidence.

Suggestions for Applying Ideas

Try something very new in your professional or personal life

▶ Think of one or more potentially valuable tasks or actions that you have long wanted to take on or do but, frankly, did not act on because of fear. Possible examples drawn from professional work environments are mentoring a younger staff person, speaking at a major conference, managing a large project, and leading a department.

Do not follow where the path may lead. Go instead where there is no path and leave a trail.

Anonymous

▶ Informed by Aristotle's balanced definition of courage and his call for commitment, develop an action plan to do that which you have, until now, feared to do.

▶ Implement your plan in a step-by-step manner.

Read the following related lessons

▶ Lesson 9, "Go Out on a Limb"

▶ Lesson 13, "Afraid of Dying, or Not Having Lived?"

▶ Lesson 52, "Looking Ahead: Can You Spare a Paradigm?"

Study the following source cited in this lesson

▶ Aristotle. 1987. *The Nicomachean ethics*. Trans. by D. Ross and revised by J.L. Ackrill and J.O. Urmson. Oxford: Oxford University Press.

To sin by silence, when they should protest makes cowards of men.

Abraham Lincoln

8

Thank Our 50 Stars

*True patriotism is not manifested in short, frenzied bursts of emotion.
It is the tranquil, steady dedication of a lifetime.*

Adlai Stevenson

*T*he topic of and title for this lesson came out of the blue as I unwrapped our new flag on July 4, 2001. The blue field of our old flag was faded; the blue on the new one formed a vivid background for the 50 bright white stars. This lesson was written later that day, while the out-of-the-blue thoughts about the blue field were still fresh.

Thomas Jefferson, third U.S. President, who died on July Fourth fifty years after public announcement of the Declaration of Independence, wrote this in that declaration:

> We hold these truths to be self-evident, that all men are created equal, that they are endowed by their Creator with certain unalienable rights, that among these are life, liberty, and the pursuit of happiness.

Clearly, the Declaration of Independence speaks to the independence of our nation. But as explicitly stated in Jefferson's words, it also speaks to the independence and resulting infinite possibilities of U.S. citizens. Jefferson continued this theme when, in 1791, he offered this advice to a family member: "You live in a country where talents, learning and honesty are so much called for that every man who possesses these may be what he pleases" (Kaminski, 1994).

Engineering practice is challenging, especially for those of us who strive to provide competent technical services and also assume managing and leading roles. Our successes are often sprinkled with and sometimes swamped by problems, obstacles, setbacks, failures, disappointments, betrayals, and insults. We occasionally get discouraged and consider giving up, or we look with envy at other professions which, from our side of the fence, look "greener."

That's when we should stop, step back, and "thank our 50 stars." Regardless of our personal difficulties, and recognizing the social, economic, political, and environmental problems facing the U.S., there is no better place on earth. As U.S. citizens and engineers, we build on the firmest foundation in the world. Our country is the most potent launch pad for members of our profession, whether our engineering efforts are in the U.S. or elsewhere.

> *In the United States, to an unprecedented degree, the individual's social role has come to be determined not by who he is but by what he can accomplish.*
> John W. Gardner

Think of the many ways most of us engineers have lived *life*, used our *liberty*, and pursued *happiness*. We freely chose our profession, where and how we studied for it, and our previous and current employers. If any aspect of our work is unacceptable, we are at liberty to seek a remedy, or move on to another employer, or start our own business. We are free to continue our formal education, develop new methodologies, change roles within engineering, or enter other professions. When faced with ethical situations, we are free to take the high road or low road. When confronted with challenges, we are at liberty to rise to the occasion or walk away. We choose were we live, where and when we travel.

The only choice we weren't at liberty to make and, in effect, the most important choice of all, is where we were born, what country we would call home. That choice was made for us and, if the country was the U.S., our lives should exemplify our thankfulness.

Shortly after the Velvet Revolution in what is now the Czech Republic, I attended an international conference in Europe. After introducing myself to two young Czech engineers, I immediately noticed their enthusiasm and happiness and asked them about it. Basically, they said professional life was now great. They were freer in many ways, including having the liberty to communicate with fellow professionals in other countries, and now they could even travel to other nations. These are, and always have been, givens for us.

> *They live in a country that offers more freedom of choice than any in the world, yet they live like prisoners, trapped by circumstances.*
> Hal Urban

And then, in spite of and in juxtaposition with the preceding envious environment of freedom and opportunity, we inevitably have the whiners, cynics, grumblers, complainers, talkers, malcontents, and losers. "Did you hear what they're going to do now?" "What a dumb process." "This is a lousy place to work." "If I were in charge." "You know what we ought to do?" "Ain't it terrible?" They just don't get it. They are oblivious to the envious position of liberty they hold, or they see it but lack the courage, self-discipline, and ability to act constructively. Either way, their way is talk, not action, and negative talk at that. I know; at times, I've been there.

A small U.S. flag stands in my office. When success comes, the flag keeps things in perspective by reminding me of the very advantageous stage on which the success played out. And when problems occur, the flag reminds me of the many avenues open to me, as a U.S. citizen and engineer, to solve those problems. Either way, I thank our 50 stars.

Suggestions for Applying Ideas

Read the following related lessons

▸ Lesson 9, "Go Out on a Limb"

▸ Lesson 13, "Afraid of Dying, or Not Having Lived?"

Study the following source cited in this lesson

▸ Kaminski, J.P. 1994. *Citizen Jefferson: the wit and wisdom of an American sage.* Madison, Wis.: Madison House Publishers.

Refer to the following supplemental source

▸ Commission on the Advancement of Women and Minorities in Science, Engineering and Technology. 2001. "Land of plenty: diversity as America's competitive edge in science, engineering and technology." *Leadership and Management in Engineering* (Oct.), pp. 27-30.

Argues that the U.S. could be even stronger in science, engineering, and technology if it did a better job of preparing citizens from all population groups.

If we as Americans show the same courage and common sense
that motivated the men who sat at Philadelphia and
gave us the Declaration of Independence and later
the Constitution of the United States, there is no domestic problem
we cannot solve and there is no foreign foe we need ever fear.

William F. Knowland

9

Go Out on a Limb

Don't be afraid to go out on a limb. That's where the fruit is.

Anonymous

*I*n our professional lives, we engineers and others technical professionals tend to be prudent, systematic, and risk-averse. Given the great impact of much of our work on public safety, health, and welfare of the public, our overriding caution is warranted. However, we should occasionally do something impulsive, unplanned, and risky. Take a flyer. Look over the top of our silo. Leap before we look. Take a shot in the dark. Think outside of the box.

The greatest mistake you can make in life is to be continually fearing you will make one.

Elbert Hubbard

Why? Because by doing so, we are likely to discover something new and valuable about ourselves while opening doors of opportunity for ourselves, others, and our organizations. Francis Bacon, the English philosopher and statesman, put it this way: "A wise man will make more opportunities than he finds." Going out on a limb is a powerful means of encountering opportunities. Even if we do not find an opportunity, we will find out more about ourselves.

If you are introverted and shy, remember that many professional actors share this trait. However, they have learned an important principle: "Just say the lines." You can apply this advice.

Our life is in play, the clock is running. There are no time-outs, although there may be a two-minute warning. If life is becoming humdrum and work is getting boring, maybe we need to try a completely different ploy, a risky plan, perhaps a "Hail Mary." The status quo, or minor variations on it, won't do. As someone wisely, but anonymously, said, "If you do what you always did, you'll get what you always got."

Let's be our sane and safe selves most of the time. But, every now and then, go out on a limb; that's where the best fruit may be.

Suggestions for Applying Ideas

"Go out on a limb" in your professional, community, or personal life

Because going out on a limb is so individual and circumstance-specific, meaningful examples are difficult to present. However, for illustration purposes, consider the following, all of which are based on actual situations:

> *Security is a false god;*
> *Begin making sacrifices*
> *to it and you are lost.*
> Paul Bowles

▶ Contact your local or regional newspaper, television, or radio station. Explain that you are an expert in XYZ and are available to be of assistance to reporters. Nothing may come of your offer, but, on the other hand, you may become the regional XYZ expert!

▶ Think of a prestigious or otherwise desirable organization that you would like to serve as a consultant or an employee. Call a decision-maker at that organization and tell them of your desire to serve or be employed and why. Nothing may happen, but, on the other hand, you may land an exciting new client or new employer.

▶ Identify a controversial issue in your field about which you have a strong, informed opinion. Submit a thought-provoking letter to the editor of a widely read journal, magazine, or newspaper. Nothing may happen, but, on the other hand, a series of enlightening follow-up letters may appear and you may meet some interesting and powerful people. You may be surprised with the number and variety of people who "come out of the woodwork" as a result of publication of your views.

▶ Pick up a newsletter/magazine/journal completely outside of your professional field. As you read it, look for articles or news items that have apparent connections or overlaps to what you do or about which you know a lot. Contact the authors/editors and note the similarities between their work and yours. Nothing may come of it, but, on the other hand, you may gain new insight and expand your network.

> *Things may come to*
> *those who wait,*
> *but only the things left*
> *by those who hustle.*
> Abraham Lincoln

▶ Approach a person in your field whom you sincerely believe is highly regarded and is powerful. Explain that you are starting out or are striving to advance and would value their advice. Depending on the physical and other circumstances, offer to "buy lunch," telephone them at their convenience, or communicate by e-mail. Be prepared with a list of five to ten concerns you have or questions you'd like to ask. Nothing may happen, but, on the other hand, you may acquire a valuable coach or even mentor.

▶ If you have basic teaching ability and desire, and possess a body of knowledge and experience to share, contact the continuing education/training department of an appropriate professional/business/education organization and offer your services (gratis or for a fee plus expenses). Going farther out on a limb, plan, advertise, conduct, and follow up on your own seminar/workshop at a hotel/conference center. Nothing may happen, but, on the other hand, you may have successful seminars and make many valuable contacts. The "worst" outcome is that you and possibly others incur expenses and the event is cancelled for lack of registrants, but even then your name and services have been exposed to hundreds or thousands of people, many of whom are potential clients, business partners, or employers.

> *Those who take bold chances don't think failure is the opposite of success. They believe that complacency is.*
> Richard Farson

▶ Call a political or other leader in your community or state. Offer your expertise for service on a committee or board. Nothing may happen, but, on the other hand, you may be asked to participate in or lead an exciting new venture.

Read the following related lessons:

> ▶ Lesson 7, "Courage: Real and Counterfeit"
> ▶ Lesson 13, "Afraid of Dying, or Not Having Lived?"

> *There is a tide in the affairs of men, which, taken at the flood, leads onto fortune; omitted, all the voyage of their life is bound in the shallows and miseries. On such a full sea are we now afloat; and we must take the current when it serves or lose our ventures.*
> William Shakespeare

10

Keeping Our Personal Financial Score

Plan ahead: It wasn't raining when Noah built the ark.
Richard Cushing

Most of us who participate in sports like running, tennis, and golf keep score. We do this partly to determine who won the particular race, match, or round. From a broader perspective, especially if we participate in the sport over a period of years, we keep score to determine our progress. That is, are we improving, have we plateaued, or are we sliding backward?

In a similar fashion, keeping score applies to our personal (or family) financial "game." That is, how have we performed financially, and what does past performance predict about our future performance?

We may be tempted to use annual earnings as a surrogate measure of our personal financial score. For example, you note that your annual income this year is 10% more than last year and conclude that you are doing just fine. I am reminded of an acquaintance who said, "Expenses rise to meet anticipated income." Given that most of us have the ability to spend in an almost unlimited fashion, using annual income as a gauge of personal financial success leaves a lot to be desired.

> *Your income does not determine your outcome.*
> Charles J. Givens

A much better measure is net worth, which is defined as assets minus liabilities. The tool for determining our net worth is the balance sheet. If we own or owe anything, we have the basis for constructing a balance sheet.

Consider using spreadsheet software for creating your balance sheet. List, in rows, all of your assets (e.g., value of residence, mutual funds, retirement accounts) and all your liabilities (e.g., balance due on residence, credit cards, college loans). Use columns to represent the end of each year. Calculate total assets, total liabilities, and net worth by year. Update the balance sheet at least annually and save the data from previous years so that you build a personal financial history.

We will derive at least two benefits from maintaining a personal balance sheet. First, it provides ready access to the asset and liability data required by banks and other lending institutions in support of applications for home mortgages, automobile loans, and other common financial transactions. Second, and this is the principal point of this lesson, we can track the "score" of our personal financial "game." Are we building net worth, plateauing, or sliding backward?

Even if we are building net worth, are we doing it fast enough to meet the financial needs consistent with our retirement and other goals? For example, assume you and your partner are each 30 years old and have a total annual income of $70,000. You want to retire completely at age 62 and, at that time, have an annual income equal to three times the buying power of today's income. Further, assume that inflation will average 3% between now and then, and you plan to live off of your net worth at age 62 by means of it yielding and growing at a rate of 7% per year. That is, you want your net worth to remain constant after age 62. Given the preceding assumptions, you will need to accumulate, by age 62, a net worth of $7.7 million.

> *He who will not economize will have to agonize.*
> Confucius

Is your net worth increasing? Is the rate sufficient to provide the financial resources to support your retirement and other goals? Keeping your personal financial "score" will help you answer these questions and, more importantly, take necessary corrective actions. Keeping our financial score puts us in the driver's seat. As succinctly stated by investment banker Roger W. Babson, "Most people should learn to tell their dollars where to go instead of asking them where they went." We must steer our dollars rather than allow them to steer us.

Suggestions for Applying Ideas

> *You can create far more wealth by how you use the money you already earn than you can from earning more.*
> Charles J. Givens

Create a personal balance sheet
▶ Possibly using the format in Figure 10-1 (Stanley and Danko, 1996)

Forecast the net worth you will need at retirement
▶ Possibly using the procedure in Figure 10-2.

	Value ($)		
	This Year	*Last Year*	*Etc.*
ASSETS			
House/condominium			
Vehicle(s)			
Stocks/bonds			
Retirement accounts (e.g., IRAs)			
Money market			
Checking account			
Insurance cash value			
Other			
Total:			
LIABILITIES			
Mortgages			
Other loans			
Credit cards			
Other			
Total:			
NET WORTH			
INCREASE IN NET WORTH (%)			

Figure 10-1 *Sample personal balance sheet.*

Compare your financial-related characteristics to the following "seven common denominators among those who successfully build wealth"

This list is based on a study of American millionaires by author, lecturer, and researcher Thomas J. Stanley and marketing professor William D. Danko (1996):

- They live well below their means.
- They allocate their time, energy, and money efficiently, in ways conducive to building wealth.
- They believe financial independence is more important than displaying high social status.
- Their parents did not provide economic outpatient care.
- Their adult children are economically self-sufficient.
- They are proficient in targeting market opportunities.

Factor	Example	You
Current age	30	
Current income ($)	60,000	
Desired retirement age	62	
Years until retirement	32	
Expected annual inflation (%)	3	
Current income adjusted for inflation at retirement ($)	$(60,000)(1.03)^{32}$ = 154,500	
Desired ratio of buying power at retirement to buying power now	2	
Income required at retirement ($)	$(154,505)(2)$ = 309,000	
Annual yield of net worth at retirement (%)	6	
NET WORTH REQUIRED AT RETIREMENT ($)	$(309,000)/(0.06)$ = $5,150,000	

Figure 16-2 *Sample personal net worth statement.*

- They chose the right occupation.

Consistent with this lesson, the authors of the cited study, precisely define wealth as follows:

> Wealth is not the same as income. If you make a good income each year and spend it all, you are not getting wealthier. You are just living high. Wealth is what you accumulate, not what you spend.

For personal finance purposes, use the engineering or decision economics you may have studied in college and/or used in your professional work (Walesh, 2000)

▶ Review the single payment compound amount factor, the single payment present worth factor, the compound amount factor, the series sinking-fund factor, the series present work factor, and other similar factors.

▶ Engineering or decision economics provides a suite of powerful personal finance tools. Concepts and factors used in decision economics are the same as those used in banking and other areas of business and personal monetary affairs.

▶ You, in effect, probably have the tools needed to make prudent personal finance decisions, such as choosing between two investment options or deciding whether or not to refinance a mortgage. These tools also enable you to verify finance terms determined by others. Use these tools to help you raise your personal financial score.

Read the following related lessons

▶ Lesson 2, "Roles—Then Goals"

▶ Lesson 3, "Smart Goals"

▶ Lesson 13, "Afraid of Dying, or Not Having Lived?"

▶ Lesson 52, "Looking Ahead: Can You Spare a Paradigm?"

Study one or more of the following sources cited in this lesson

▶ Stanley, T.J., and W.D. Danko. 1996. *The millionaire next door.* New York: Pocket Books.

▶ Walesh, S.G. 2000. *Engineering your future: the non-technical side of professional practice in engineering and other technical fields*, 2nd Ed. Reston, Va.: ASCE Press. (Chapter 8, "Decision Economics," and Chapter 10, "Business Accounting Methods.")

Refer to one or more of the following supplemental sources

▶ Belsky, G., and T. Gilovich. 1999. *Why smart people make big money mistakes—and how to correct them.* New York: Simon & Schuster.

Motivates us to think about behavioral economics, which explains why we sometimes make illogical decisions that adversely affect our financial well-being. For example, is one of the following two decisions more prudent and, if so, which one: saving $25 on a $200 purchase or saving $25 on a $2,000 purchase? Really? Read this book to learn the answer and to learn more about useful topics, including the power of compounding, accounting for inflation, the sunk cost fallacy, and mental accounting.

▶ Handy, C. 1998. *The hungry spirit: beyond capitalism: a quest for purpose in the modern world.* New York: Broadway Books.

Recognizes the importance of financial well-being but notes that "money is the means of life and not the point of it." Advocates "proper selfishness," which is defined as accepting "responsibility for making the most of oneself by, ultimately, finding a purpose beyond and bigger than oneself."

> *Almost any man knows how to earn money, but not one in a million knows how to spend it.*
> Henry David Thoreau

▶ Kiyosaki, R.T., with S.L. Lechter. 1998. *Rich dad, poor dad: what the rich teach their kids about money—that the poor and middle class do not!* New York: Warner Books.

Stresses the importance of financial literacy, notes that it is not taught in school, and stresses the need to teach and learn financial literacy at home and/or on our own. Numerous asset-building ideas and tips are offered.

Subscribe to one or more of these e-newsletters

▶ "Money Matters," provided free by Quicken, offers timely ideas and information to help individuals manage their money. This monthly e-newsletter typically discusses investments, retirement, savings, loans, and taxes. To subscribe, go to http://quicken.com.

▶ "TaxTalk," provided free each month by the National Association for the Self-Employed (NASE). Offers practical advice on a wide variety of tax issues such as how to correct an error on a tax return, determine the status of a refund, and estimate social security benefits. To subscribe, go to http://www.nase.org.

Visit one or more of these websites

▶ "Kiplinger.com" (http://www.kiplinger.com) is the website of The Kiplinger Washington Editors, Inc. Included are calculators to assist with financial planning and decisions, such as determining net worth, estimating the cost of raising a child, and saving for college.

▶ "MSN Money" (http://moneycentral.msn.com/home.asp) is maintained by CNBC. Includes practical advice on the elements of financial planning, such as retirement needs, debt management, and insurance.

▶ "Quicken" (http://www.quicken.com) is maintained by Intuit. Included are an extensive glossary and stock market updates.

▶ "Smartmoney" (http://www.smartmoney.com/pf/) is the website of *Smart Money* magazine. Includes financial planning worksheets and calculators.

Some know the price of everything and the value of nothing.

Anonymous

Job Security Is an Oxymoron, Career Security Doesn't Have to Be

I must create a system or be enslaved by another man's.
William Blake

About 35 years ago, I called my mother to tell her I was leaving my first real job. She was dismayed; she thought her eldest had been let go, fired! I assured her I was leaving by my choice; I had done what I could do and wanted to move on. My mother lived through the hard times of the Great Depression when, if you were lucky enough to have a job, you did everything you could to keep it. My mother knew the importance of a secure job.

A decade after the Depression, the U.S. was in World War II, and a decade later our country was a global economic leader (Goodwin, 1994). The secure jobs my mother valued had become commonplace. Workers and employers entered into informal agreements whereby workers gave a day's work for a day's pay and the employer provided a secure job. Workers often received gold watches after 25 years of service.

If money is your hope for independence you will never have it. The only real security that a man can have in this world is a reserve of knowledge, experience and ability.
Henry Ford

Today gold watches, and the job security they represented, are rare. Job security, which may be defined as knowing you will be allowed to do the same thing at the same place for a long time, is an oxymoron. Job security passed away because of increased personal productivity, global competition, outsourcing, privatization, reengineering, and consolidation of organizations.

However, there is good news! Enlightened technical professionals who carefully manage their careers can thrive in the new world of work. They can enjoy career security, that is, knowing they will always be employable somewhere, and have fun, success, and satisfaction in the process. Guaranteed employment with one organization is out; guaranteed employability is a viable replacement (Handy, 1998; Walesh, 1997a).

Thriving in today's economy requires each of us, whether currently an employee or a freelancer, to think, plan, and perform as though we were on our own, a free agent, or an independent economic unit. Offered here are six suggestions for earning career security as an engineer in the modern work world. Each career security suggestion is developed further in the applications portion of this lesson.

▶ *Protect your reputation.* An engineer usually provides advice in the form of a report or a design, not a material product. The credibility imparted to that advice is closely tied to the professional's reputation. The client usually is not able to judge the quality of the advice. However, the client is quite capable of judging the quality of the professional based on facts or perceptions.

▶ *Attend to personal and professional development.* Another aspect of thriving in today's professional market is creating and implementing an ambitious personal and professional development plan. Some employers expect us to plan our personal and professional development and help us with it; however, such employers are rare. Clearly, in the face of such organizational indifference, you must take the lead. Don't expect your employer to "take care of" your development.

> *Security isn't securities. It's knowing that someone cares whether you are or cease to be.*
> Malcolm Forbes

▶ *Invest a dime of every dollar.* Invest at least 10% of your earnings, beginning with the first dollar on your first job and extending throughout your career. Leverage your investments by taking full advantage of employer- and government-sponsored programs that match your investments and shield your investments, and the earnings on them, from income taxes.

▶ *Enhance communication competence.* Effective communication is necessary, although not sufficient, to realizing your potential. Unless you are a genius, are inextricably linked to business ownership, or enjoy some other rare privilege, you will need effective communication skills to earn career security. Listening, speaking, writing, use of visuals, and application of mathematics are the five pivotal communication skills needed by the technical professional.

▶ *Cultivate contacts.* The current term for cultivating contacts is networking. Only about one-fifth of all available engineering positions will appear in newspapers and other written announcements. Most position openings are communicated informally and, therefore, known only to those who are members of appropriate networks.

▶ *Hang out a shingle.* Start a part-time business. Choose something that you are good at and that is not necessarily related to your principal profession. The product or service you offer might be linked to a hobby or special interest. Perhaps you can teach tennis, form a musical group, write poetry, or paint houses. Do this to deepen and broaden your understanding of business.

The demise of job security is a positive development for the engineering profession. It is stimulating growth of a subgroup of independent, entrepreneurial, "on top" rather than "on tap" engineers and other technical professionals who have earned career security. We can join that group by practicing an enlightened selfishness intended to optimize our potential to be faithful stewards of the gifts we've been given. There are only two futures for any of us: the one we create for ourselves, or the one others create for us.

Suggestions for Applying Ideas

Consider these thoughts on protecting our reputations

▶ Our reputation is what colleagues, clients, and others really "hear" and "see" when we speak, write, or otherwise interact with them. Guard that reputation as though it were your most valued asset; it probably is.

▶ Consider two important facets of one's reputation: honesty and integrity.
  • We tend to indiscriminately lump them together in a kind of moral mush.
  • However, speaker and author Stephen Covey (1990) offers some discriminating guidance. He says, "Honesty is telling the truth—in other words conforming words to reality." He goes on to say, "Integrity is conforming reality to our words, in other words, keeping promises and fulfilling expectations." Stated differently, honesty is retrospective and integrity is prospective.
  • Honesty is what we say about what we've done, and integrity is what we do about what we've said.

▶ Old-fashioned as the following advice may sound, it is offered with all sincerity: Tell the truth, keep your word, do your share of the work, give credit where credit is due, and don't blame others.

▶ The behavioral expectations of the engineering profession are more specific and more demanding than society at large. For example, in working with clients, we acquire considerable information about them and their operations. Sharing such information with other clients and other contacts, although it might appear helpful, is blatantly unethical and, therefore, potentially damaging to our reputations. Protecting your reputation within business and professional circles requires a working familiarity with ethics codes established by the engineering profession. Engineering codes, such as those of the National Society of Professional Engineers (NSPE), the American Society of Civil Engineers (ASCE), the American Society of Mechanical Engineers (ASME), and the Institute of Electrical and Electronics Engineers (IEEE), contain strong provisions that provide client confidentiality. For example, the NSPE code (2003) states:

> Engineers shall not disclose, without consent, confidential information concerning business affairs or technical processes of any present or former client or employer, or public body on which they serve.

▶ Good intentions are not enough. Naivete can do you in. Ignorance is no excuse. We should know what is expected within the engineering profession.

► Years ago, my wife and I purchased a crystal wine decanter as a memento of a visit to Munich. Each person's reputation is like a handcrafted crystal piece.
  • Like the crystal, each person's reputation is unique.
  • Like the crystal, each person's reputation has many facets.
  • Like the crystal, a long time is required to create a reputation.
  • Like the crystal, once damaged, a reputation may never be repaired.

► Finally, good news about people travels fast; bad news travels even faster.

Apply some of the following approaches to personal and professional development

► "Personal and professional development can include developing understanding of and competence in goal-setting, personal time management, communication, delegation, personality types, networking, leadership, the sociopolitical process, and effecting change" (ASCE, 2003).

► "In addition to the preceding, professional development can include career management, increasing discipline knowledge, understanding business fundamentals, contributing to the profession, self-employment, additional graduate studies, and achieving licensure and specialty certification" (ASCE, 2002).

► Take charge of your personal development by using these three learning mechanisms available to you as a practicing professional:
  • Just-in-time on the job learning
  • Classes, seminars, and workshops
  • Active involvement in professional societies

► A mentor can enhance the content and accelerate the implementation of your personal and professional development plan. The mentor should be someone you know and trust and preferably not be your boss. Select someone who will respect confidences and offer constructive criticism and advice. A mentor can be a sounding board and perhaps your safety net.

► Holding a current engineering license is one indication that you have probably completed a demanding formal education and acquired basic experience. Furthermore, because of ever more demanding licensure laws, licensure indicates that you are maintaining at least minimal competence through continuing education. While licensure provides an edge in essentially all engineering disciplines, career security in civil engineering is severely hampered without one. ASCE Policy Statement 465, adopted in October 2001, stipulates licensure as one requirement of being a civil engineer practicing at the professional level (ASCE, 2003). If you intend to practice engineering, earning an engineering degree and not becoming licensed is like buying a sports car and leaving it in the garage.

Consider these ideas on investing for the future

► Because of the power of compound interest and the long-term appreciation of stocks, you can build a large retirement fund. Furthermore, given the vagaries of the employment market, prudent investing also will quickly accumulate a financial safety net.

▶ You must start investing early, and if you haven't started yet, then start now. Avoid this trap: "Wow, that dime of every dollar idea is great! As soon as I can afford it, I'm going to do it."

▶ Project retirement needs and monitor progress toward meeting those needs.

▶ Consider retaining the services of a financial advisor to help you decide how much to invest and how to allocate your investment advisor. A typical annual cost of such services is 1% of the assets managed.

Apply the approaches described elsewhere in this book to enhance communication competence

▶ Depending on your communication strengths and weaknesses, selectively study Lessons 14 through 24 in Part 2, "Communication," of this book. (Note that, after Part 1, "Personal Roles, Goals, and Development," Part 2, "Communication," is the second largest section of this book. And rightly so, given the critical role communication ability plays in personal success.)

Consider these thoughts on the importance of networking

▶ If you are or someday might seek employment, network, network, network! That's essentially all of us. As a former engineering manager, I conducted numerous personnel searches. We avoided newspaper and similar ads whenever possible because, no matter how tightly written, they attracted too many unqualified candidates which, in turn, required considerable time to process. We focused, instead, on people we knew or knew of, or who were referred to us.

▶ One of the best ways to network is to become actively involved in a few carefully selected professional and community organizations. The emphasis here is on active involvement, rather than passive membership. If you focus on making contributions, contacts will occur and good things will happen. Why? Because you will be known as a competent and contributing individual. People like that make great employees.

▶ Do not join a professional or community organization primarily to make contacts. Your motive will be obvious to most, especially if you contribute little or nothing.

▶ An important aspect of networking and cultivating contacts is carefully choosing your teachers, supervisors, business partners, and colleagues. These associations are especially important early in your career because they deeply influence your attitude about and approach to professional work. Strive to associate with ethical, creative, entrepreneurial individuals.

Reflect on these aspects of hanging out a shingle

The expression "hang out a shingle" is derived from the tradition of hanging a small sign, or "shingle," outside the professional's office and listing his or her name and profession.

► A part-time business will broaden your awareness and cause you to develop more non-technical knowledge and skills in areas such as marketing, accounting, interpersonal communication, tax laws, ethics, liability, and retirement planning.

► A part-time business will help you develop as a manager and leader. Deciding what to do and how to do it will be your responsibility.

► Probably most important, a part-time business will demonstrate how the economic value of knowledge, a product, or a service lies in how it meets client or customer needs.

► And who knows, you might earn a supplemental income.

► A word of caution: Starting a part-time business while the employee of an organization may involve ethical issues. For example, the ASCE Code of Ethics (2002) states:

> Engineers shall not accept professional employment outside of their regular work or interest without the knowledge of their employers.

► See Lesson 12, "Are You Unemployable?" because success with a part-time business may stimulate thinking about your own full-time business.

Read the following related lessons

► Lesson 10, "Keeping Our Personal Financial Score"

► Lesson 12, "Are You Unemployable?"

► Lessons 14 through 24 in Part 2, "Communication"

► Lesson 25, "Professional Students"

► Lesson 47, "Eagles and Turkeys"

Study the following sources cited in this lesson

► ASCE. 2003. *Civil engineering body of knowledge for the 21st century: preparing the civil engineer for the future.* Reston, Va.: ASCE. (Available at http://www.asce.org/raisethebar.)

► ASCE. 2002. *Official Register 2002.* Reston, Va: ASCE. (Available at http://www.asce.org/inside/codeofethics.cfm.)

► Covey, S.R. 1990. *The 7 habits of highly effective people: restoring the character ethic.* New York: Simon & Schuster.

▶ Goodwin, D.K. 1994. *No ordinary time.* New York: Simon & Schuster.

Gives a fascinating account of U.S. growth in military and industrial might during World War II.

▶ Handy, C. 1998. *The hungry spirit: beyond capitalism: a quest for purpose in the modern world.* New York: Broadway Books, pp. 61-67.

Provides an additional, but somewhat negative, discussion of loss of guaranteed employment and the loss of guaranteed employability.

▶ National Society of Professional Engineers. Code of Ethics for Engineers. (Available at http://www.nspe.org/ethics/eh1-code.asp.)

▶ Walesh, S.G. 1997a. "Job security is an oxymoron." *Civil Engineering* (Feb.), pp. 62-63.

Describes a positive approach to loss of job security because of the potential for career security.

The psychological contract between employers and employees has changed.
The smart jargon now talks of guaranteeing "employability,"
not "employment," which . . . means don't count on us,
count on yourself, but we'll try to help if we can.

Charles Handy

12

Are You Unemployable?

*The greatest of all human benefits, that, at least, without which
no other benefit can be truly enjoyed, is independence.*

Parke Godwin

Are you, a once reliable, accomplished, and regularly employed professional, increasingly finding working for someone else intolerable? Are you
becoming unemployable? Are you beginning to think about going out on
your own, flying solo as a sole proprietor or freelance consultant? Consider
the following common motivators for flying solo (Walesh, 2002).

▶ *Career Security.* Job security, as symbolized by receiving a gold watch after
many years of employment with one company, is history. While job security
is an oxymoron, career security can be a reality. In today's economy, the individual consultant who serves several clients at any one time often has more
security than the full-time employee of an organization.

▶ *Bureaucratic Excess.* The more you monitor the situation, the more you
see bureaucratic excess. One symptom is cumbersome decision-making, and
another is being spread too thin. You are asked to do more and more with less
and less. Will you continue to do less and less on more and more until you are
eventually doing absolutely nothing on everything? Another
bureaucratic excess is rampant reporting on what you are
doing or trying to do, which leaves even less time for doing
productive work.

*You can be young
without money, but you
can't be old without it.*

Tennessee Williams

▶ *Lagging Net Worth.* You worry that your earnings may be
lagging. Your net worth is growing too slowly to ensure a comfortable retirement. Even though you are prudent, you won't
make the necessary net worth unless you significantly increase your income.
If you're going to work as hard as you do, you might as well work for yourself
and, if successful, reap the monetary benefits.

▶ *Stagnation.* Stagnation means plateauing in terms of knowledge and skills. You recognize a growing gap between your desire for challenge and your autonomy to choose challenging assignments. Perhaps you want to stretch yourself technically or further develop the manager and leader in you. Maybe you need to explore options that will once again enable you to unilaterally take on technical and non-technical challenges, gain in knowledge and skills, and be productive and happier in the process.

▶ *Shackled by Success.* Perhaps you've made significant, high-profile contributions to your organization in a relatively narrow area. For example, you've worked hard to become a regionally or nationally recognized expert in some aspect of your business. Frankly, you're tired of this and want to take on completely new challenges. But you are perceived by the principals of your firm as being able to do only one type of specialty work. You've repeatedly told them about your other goals, but the message doesn't register. You are their guru and are shackled by your success.

▶ *Pink-Slipped.* Being dismissed by your employer, especially unexpectedly, is truly a "sign" that one needs to explore options. While being bypassed or demoted are experiences that are likely to be less traumatic than being dismissed, they also can have serious consequences, such as bruised ego and denied income. The dismissed, bypassed, or demoted person is likely to go through a series of strong, mostly negative feelings, such as anger, anxiety, depression, despair, devaluation, failure, fear, inadequacy, jealousy, loss, mourning, sadness, self-doubt, shame, shock, and surprise. Near the end of this process, many bypassed, dismissed, or demoted professionals will be exploring options. One option is to become an individual practitioner. Maybe this disastrous bypassing, dismissal, or demotion can be turned into an uplifting opportunity.

> *We are socialized to believe that endless luxury and leisure are what will make us happy, but it is not true. Endless leisure leads to apathy and despair. What feels good to most people is being useful. What gives life meaning is its effects on other people. Rest feels good only if it is in contrast to work.*
>
> Mary Pipher

▶ *Still Have the Itch.* So, you finally retired! You accomplished much professionally and have many pleasant memories (and a few not-so-pleasant memories). Frankly, you are happy to have most of it behind you. But you still have the itch. There's some unfinished business and it has to do with business. You miss certain aspects of working, including usefulness and meaning. Perhaps you can re-enter the business and professional world as a freelancer, largely on your terms. Maybe you can work on what you want, when and where you want.

Look before you leap! Before voluntarily leaving your current employer to fly solo, take one more look at internal options. This advice is especially important if you are employed in a large, complex, and multi-office public or private organization. Do a reality check by asking yourself these questions:

- Do the appropriate decision-makers know of your professional desires and frustrations?

Keep away from people who try to belittle your ambitions. Small people always do that, but the really great make you feel that you, too, can become great.

Mark Twain

- Might an organizational change be imminent (e.g., acquisition, revamped structure, additional service lines, retirements, or new office) that would fit your need?
- Are your expectations realistic? While conventional thinking should not be your guide, you should determine if your expectations are in the realm of reality.

Assuming that one or more of the seven listed motivators apply to you (and perhaps there are others), and assuming there are no suitable opportunities with your current employer, then you might be ready to "go for it" or, in the more eloquent words of essayist and poet Henry David Thoreau, "Go confidently in the direction of your dreams. Live the life you imagined." The next section can help you determine your readiness to be an independent consultant.

Suggestions for Applying Ideas

Evaluate yourself against the following factors that point to success as an independent consultant (Walesh, 2002)

▶ *Inquisitiveness and currency of knowledge:* The consultant is often retained to provide expertise the client does not possess. Consultants, as individuals or as organizations, should define their areas of expertise and remain current in them. On the surface, consultants seem successful primarily because of the answers they provide based on their knowledge and experience. However, the questions they ask their clients, others, and themselves are more important than the answers they give. Once key questions are asked, the consultant knows how to find the answers. *Do you know how to find answers?*

▶ *Responsiveness to schedules and other needs:* The consultant may be retained because the client does not have the personnel to complete a task or do a project. If the effort is late because of the consultant, the principal reason for retaining the consultant is negated. Responsiveness to client needs and schedules requires that the consultant have a strong service orientation. *Are you service oriented?*

▶ *Strong people orientation:* Although engineers plan, design, construct, fabricate, manufacture, and care for "things," they are doing this for the benefit of people. The consultant matches the needs of people with the appropriate applications of science and technology. Accordingly, effectiveness in consulting requires communication skills, with emphasis on listening, writing, and speaking. While the successful consultant enjoys interacting with people, some individuals are not very pleasant because they are under great personal or organizational stress, or perhaps it is just their basic personality. Inflating the egos of some clients and tolerating the over-inflated egos of others could take a toll on your own ego. The people challenges of consulting are further complicated by frequent changes in clients, potential clients, and their representatives. *Is your skin "thick enough"?*

▶ *Self-motivated:* Most of what consultants do for clients is at the consultant's initiative within the overall framework established by the consultant-client agreement. Clients tend to assume that if they are not hearing anything from "their consultant," the consultant is proceeding with the project in a timely fashion. Moreover, the consultant must be available, on very short notice, to answer a question, give advice, or provide a status report or other accounting of the efforts to date. *Are you predisposed to keep things on track?*

▶ *Creativity:* Consultants must have the ability to be creative, to synthesize, and to see previously unforeseen patterns and possibilities. The typical project involves technical, financial, regulatory, personnel, and other facets, all of which can be easily assembled in a variety of suboptimal ways. *Are you creative?*

▶ *Physical and emotional toughness:* The successful consultant needs physical and emotional strength to withstand pressure, long hours, and travel. Some of the consultant's meetings and presentations are very difficult because they occur in situations highly charged by personality conflicts, political pressure, financial concerns, environmental impacts, and liability issues. In addition, consultants are often not selected for projects even though they believe they were the most qualified or had the best proposal. Frequent rejection can take its toll on conscientious and competent individuals, but it is one of the realities of the consulting field. The rare highs of winning, especially early on, may not be sufficient to offset the many lows of losing. *Are you tough enough?*

Use the "hats" method (Walesh, 2002) as another way to determine if you could be successful as a sole proprietor

▶ How many "hats" do you wear now? If you are employed by a public or private engineering organization, you probably "wear" several "hats." A typical combination might be designer, project manager, and department head.

▶ As an independent consultant, you will have to wear, or attend to, many more hats. Examples are accountant, chief executive officer, coach, computer expert, creator, documenter, dreamer, facilitator, friend, integrator, lawyer, listener, marketer, mentor, partner, planner, prime contractor, project engineer, project manager, speaker, subcontractor, teacher, and writer.

▶ Because of the heavy demand on sole practitioners to "wear many hats" and attend to so many matters, some develop mutually beneficial alliances with other sole proprietors or even companies.

▶ Are you willing and able to wear many more "hats"?

▶ As a freelancer, you may be able to cast off some "hats" that never fit you well. Examples are gopher, outsider, politician, scapegoat, and "yes" man/woman.

Consider the assets you "bring to the table" (Walesh, 2002)

▶ One way to identify and evaluate your non-monetary assets is to view them as being in one of three categories:
- Your skills, knowledge, and characteristics: Who you are.
- Your network: Who you know, and more importantly, what they really know and think about you.
- Your client awareness: What you know about client needs/wants.

▶ Your business opportunity is likely to lie within the intersection of the three asset categories.

Assess your contacts realistically

▶ Carefully evaluate the transformability and resiliency of your network and of your potential clients.

▶ An example: Assume you are an employee of a government entity or a business that uses consultants you view as potential clients. Or, as a government or business employee, perhaps you view other government entities and businesses, with which you had contacts, as potential clients.

▶ Be cautious. Because you have cordial and mutually beneficial relationships with these government and business contacts does not necessarily mean they have high enough regard for your abilities to retain you.

▶ What you interpret as friendship, confidence, and admiration may simply be "schmoozing" of you by them.

▶ Bottom line: Assuming they have needs, would a significant number of your current professional colleagues think enough of you to retain you to meet those needs?

Think of ways that ideas and advice presented in this lesson might enable you to be a better employee, perhaps by developing more entrepreneurial attitudes, knowledge, and skills

▶ Could deficiencies in your personal work habits, such as poor time management and ineffective delegation, be keeping you bogged down in bureaucratic excess?

▶ Are you stagnating because you lack goals and action plans and/or are not proactive in sharing your aspirations with supervisors?

▶ Are you shackled by success, that is, viewed as a one-dimensional person because you shun opportunities to broaden your responsibilities, to wear more or different "hats"?

Read the following related lessons

▶ Lesson 3, "Smart Goals"

▶ Lesson 7, "Courage: Real and Counterfeit"

▶ Lesson 11, "Job Security Is an Oxymoron, Career Security Doesn't Have to Be"

▶ Lesson 42, "A Simple Professional Services Marketing Model"

▶ Lesson 43, "Speed as a Competitive Edge"

Study the following source cited in this lesson

▶ Walesh, S.G. 2002. *Flying solo: how to start an individual practitioner consulting business*. Valparaiso, Ind.: Hannah Publishing.

Refer to one or more of the following supplemental sources

▶ Fenske, S.M., and T.E. Fenske. 1989. "Business planning for new engineering consulting firms." *Journal of Management in Engineering* (Jan.), pp. 89-95.

States that the probability of success of a new engineering consulting firm will be greatly increased by preparation of a business plan. Essential components are a statement of purpose, a marketing plan, and pro forma financial statements.

▶ Lindeburg, M.R. 1997. *Getting started as a consulting engineer*. Belmont, Calif.: Professional Publications.

Provides an overview of topics such as forms of ownership, insurance, marketing, fees, contracts, client relations, and ethics.

▶ Oxer, J.P. 1999. "The independent contractor." *Journal of Management in Engineering* (Jan./Feb.), pp. 18-20.

Identifies advantages gained by employers through use of independent consultants. Two examples are access to specialized expertise, and only when needed, and the ease with which a relationship can be terminated. Urges firms, before they retain an independent consultant, to check references and credentials and perform a background check.

▶ Perlstein, D. 1998. *Solo success: 100 tips for becoming a $100,000-a-year freelancer*. New York: Three Rivers Press.

Offers details applicable to a wide range of sole proprietor consulting businesses, although not specifically engineering.

Visit one or more of these websites

▶ "AllBusiness" (http://www.allbusiness.com/) is the website of AllBusiness. Potentially useful features include forms and agreements and lessons learned based on experiences of entrepreneurs.

▶ "Free Agent Nation" (http://www.freeagentnation.com), produced by Daniel H. Pink, provides frequently asked questions and many free articles targeting people who work for themselves.

▶ "Nolo Law for All" (http://www.nolo.com) is maintained by Nolo, a company whose mission is "to make the legal system work for everyone—not just lawyers." Website content includes a section on independent contractors which provides many free, informative articles.

▶ "SBA Starting Your Business" (http://www.sba.gov/starting) is a website of the U.S. Small Business Administration. Included are frequently asked questions and sections on start-up basics, financing, employees and taxes, and a business plan tutorial.

▶ "Working Solo"(http://www.workingsolo.com) is maintained by Terri Lonier of SOHO (Small Office/Home Office). Includes ten frequently asked questions, a series of articles, and links to many websites.

Subscribe to one or more of these e-newsletters

▶ "TelE-Sales Tips," a product sales-oriented newsletter that focuses on question asking. Subscribe by going to http://www.BusinessByPhone.com.

▶ "Working Solo eNews," a free monthly newsletter that "brings news, information, tips and insights on self-employment." To subscribe, go to http://www.workingsolo.com/

There is a time in every man's education when he arrives at the conviction that envy is ignorance; that imitation is suicide; that he must take himself for better, for worse, as his portion; that though the wide universe is full of good, no kernel of nourishing corn can come to him but through his toil bestowed on that plot of ground which is given to him to till. The power which resides in him is new in nature, and none but he knows what that is which he can do, nor does he know until he has tried . . . God will not have his work made manifest by cowards.

Ralph Waldo Emerson

13

Afraid of Dying, or Not Having Lived?

We all live under the same sky, but we don't all have the same horizon.
Konrad Adenauer

Rabbi Harold S. Kushner, in his book *Living a Life That Matters* (2001), says, "The dying have taught me one great lesson . . . most people are not afraid of dying, they are afraid of not having lived."

We engineers have a wealth of resources to help us live full lives, to avoid having "not lived" regrets. Although not the highest, our compensation is certainly adequate. Frankly, we are generally of above-average intelligence, which was one of the reasons we were admitted to engineering colleges in the first place. Our education, experience, and can-do attitude enable us to define problems, create alternatives, compare them, make choices, and implement the desired course of action. Changing employment or finding employment after employment lapses is fairly easy, subject to some complications as we get older.

However, as Rabbi Kushner observes, we engineers have some liabilities that could easily lead to regrets later in life. We tend to be careful, which, given our responsibilities for public safety, health, and welfare, is usually an appropriate trait. However, our cautious nature may narrow our vision of what we might do with the rest of our lives. We also tend to be very logical and to rationalize; we are prone to excessive quantification and uncomfortable with ambiguity. For example, in a moment of personal brainstorming, if we were to contemplate dropping out of employment for a year and doing something very different, any one of us could probably come up with ten reasons why we couldn't or shouldn't.

Do not be too timid and squeamish about your actions. All life is an experiment.

Ralph Waldo Emerson

In the mid 1990s, my wife and I devoted six months to traveling 5,000 miles on our boat around much of the eastern United States. On return, we were invited to give several presentations. What is memorable about those presentations is that, inevitably, one or more individuals would come up to us after the presentation and tell us about some dream or fantasy they had. Based on these and other experiences, I conclude that most of us have visions of unusual things we would like to do. These ideas, if acted on, constitute energetically and creatively living our lives.

> *Make sure you smell the roses before you push the daisies.*
>
> Anonymous

Well, let's do at least some of those things. Let's avoid entering the golden years with regret for not having fully lived. Hike the Appalachian Trail, start a used book business, bicycle across the U.S., become a teacher, enter the ministry, do nothing for a year. Yes, I know there are many reasons why we can't. But there is one overwhelming reason why we should: we only go around once.

Suggestions for Applying Ideas

Grab pen and paper and start a list of things you've repeatedly dreamed of doing, of experiences you've fantasized about (this might be a joint effort with one or more loved ones or friends)

▶ For each item, ask yourself why you want to do it. "Because" is an acceptable default answer.

> *Courage is resistance to fear, mastery of fear, not absence of fear.*
>
> Mark Twain

▶ For each item, indicate why you "can't do it" and what you fear.

▶ Select one dream or fantasy item—your first choice. View the reasons you "can't do it" and the fears, not as barriers but as obstacles to be overcome or problems to be solved. Create an action plan to overcome or solve each of them.

▶ Start to implement your plan.

Avoid "but-phobia" or "but-neurosis" and practice the "other side of but" instead (Bach, 1970; Walther, 1991)

▶ "But-phobia" and "but-neurosis" refer to the tendency to make a positive statement and then immediately follow it with a negative thought, leaving on the balance a negative, pessimistic message.

▶ For example: "Traveling in Europe for three months would be a wonderful educational experience, but we could not afford it." Or: "Writing a book would be satisfying, but I don't have the time."

▶ Author Marcus Bach, in his book *The World of Serendipity* (1970), advises us to consider "the other side of but." He states that "To reverse your point of view is to start your life anew." More specifically, he suggests revisiting the way we say things to make the "but" part of our statements positive.

▶ Following this advice, the preceding negative statement about European travel could be turned around to become this positive statement: "The three-month European trip will be educational and expensive, but we will embark on a four-year savings program to generate the necessary funds."

> *If you think small, you'll stay small.*
>
> Ray Kroc

▶ The defeatist book writing expression could be turned around to become this winning expression: "The satisfying book writing experience will require a major time investment, but I will schedule one hour of writing per day for the next six weeks to launch the project."

Interject the idea of living energetically and creatively into your daily and weekly routine

▶ While you can't stop the treadmill, you can hop off for a while. For example:
- Take a different route when you drive or walk home from your office.
- Rent a convertible for your birthday, take a drive in the country with someone special, and have a picnic.
- Stop and browse in that funky gift shop you've passed hundreds of times.
- Enroll in an art class.
- Fly to London for a long weekend.
- Send a special card to your spouse, friend, or colleague, just for the heck of it.

Read the following related lessons

▶ Lesson 7, "Courage: Real and Counterfeit"

▶ Lesson 9, "Go Out On a Limb"

▶ Lesson 47, "Eagles and Turkeys"

Study one or more of the following sources cited in this lesson

▶ Kushner, H.S. 2001. *Living a life that matters.* New York: Alfred A. Knopf.

▶ Bach, M. 1970. *The world of serendipity.* Marina Del Ray, Calif.: DeVorss & Company.

▶ Walther, G.R. 1991. *Power talking: 50 ways to say what you mean and get what you want.* New York: Berkley Books.

Visit this website

▶ "Making a Life, Making a Living" (http://www.makingalife.com/) is maintained by Mark S. Albion. As suggested by the title, meaningful living is stressed. This website markets products and includes a free quote search feature.

The mind, ever the willing servant, will respond to boldness,
for boldness, in effect, is a command to deliver mental resources.

Norman Vincent Peale

Communication

"The problem was communication." How often have we heard this as the explanation or excuse for failure or a less-than-satisfactory result? The project closed in the red, the city council rejected our capital improvement program, we didn't land the customer, we lost another productive and promising employee—more often than not, poor communication is the cause. It is not an excuse hiding the real cause.

Communication, or lack thereof, does not have to cause problems. Communication can prevent them. We can turn communication as a liability into communication as an asset. How? By strengthening our communication knowledge, skills, and attitudes and by exercising the self-discipline needed to apply them.

Part 2 presents a variety of communication ideas and applications that are intended to elevate awareness of the importance of communication and to enhance your communication competence. Various communication modes are addressed, ranging from the all-important asking and listening to preparing, presenting, and publishing professional papers.

14

Communicating Five Ways

He who thinketh by the inch and talketh by the yard
should be kicketh by the foot.

Walter A. Johnson

*T*o realize our potential in our work, community, and personal lives, we must communicate effectively. The most exciting vision, the most thoughtful insight, the most elegant solution, and the most creative design are all for naught unless they are effectively communicated to others.

If you lack the ability to communicate well, the intellectual and other seeds that you plant with colleagues, clients, friends, family, and others are not likely to germinate, sprout, and bear fruit, denying everyone the bounty of your labors. Communication competence is especially important if we want to develop career security.

With rare exception, effective communication skills are vital to realizing potential and earning career security. Unless you are the non-communicative genius in the workplace or the recent technical program graduate who also happens to be the boss's daughter or son, recognition and advancement are dependent on your ability to communicate.

Engineers need to know and practice the following five forms of communication:

1. ***Mathematics:*** We engineers tend to be proficient in the use of arithmetic, geometry, algebra, and calculus in describing quantities and relationships. Mathematics is one of our most comfortable communication forms. It's usually definitive. Tobias Dantzig, mathematician and author, said "Mathematics is the supreme arbiter. From its decisions, there is no appeal."

If I cannot picture it,
I cannot understand it.

Albert Einstein

2. ***Visuals:*** Visual communication includes the graphics and props used in a presentation as well as the speaker's dress and body language.

3. *Writing:* Engineers routinely write e-mails, letters, specifications, memoranda, reports, and other documents. Those of us who want to improve our managing and leading skills should see writing as an essential function that warrants continuous improvement. William Zinsser, a writer, editor, and teacher, linked writing and leadership by noting that "Writing is the handmaiden of leadership; Abraham Lincoln and Winston Churchill rode to glory on the back of a strong declarative sentence." Management writer and speaker Thomas L. Brown recognized both the challenge and value of good writing when he said "Hard writing makes for easier reading."

4. *Speaking:* Within engineering practice, speaking takes many forms ranging from informal presentations for small groups to formal papers delivered at international conferences. Effective managing and leading requires mastery of this skill. Work on vocabulary, intonation, and accent. Unfortunately, many of us fear speaking. Journalist Roscoe Drummond, said "The mind is a wonderful thing. It starts working the minute you're born and never stops until you get up to speak in public." None of us can escape speaking, as explained by speaking consultant and author Bert Decker, who said, "...we are all public speakers. There's no such thing as a private speaker—except a person who talks to himself."

5. *Listening:* This easy-to-overlook communication channel is essential to building interpersonal relationships. When practiced at its best, listening means hearing and understanding the words as well as the feelings accompanying them. Stephen Covey (1990), speaker and author, admonishes us to "Seek first to understand, then to be understood." And the Bible (James 1:19) advises us to "be quick to hear, slow to speak, slow to anger"

> *The difference between the right word and the almost right word is the difference between lightning and the lightning bug.*
>
> Mark Twain

In general, the list reflects the order of importance the items receive in engineering education. We should not be surprised that many young engineers, and some older ones, need the most work in listening and speaking—they have had very little instruction and practice in either.

The complete communicator is competent in all five forms of communication and excellent in some. Through a judicious blend of self-discipline and practice, we can continue to improve our communication ability, as well as the quality of our lives and of those around us.

Suggestions for Applying Ideas

Take action to enhance your communication competence

▶ Offer to take notes at a meeting, thereby strengthening your observing, listening, and writing skills.

▶ Ask to help make a project presentation to a client, thereby enhancing your speaking and visual skills.

▶ Join Toastmasters International (http://www.toastmasters.org/), and develop your own speaking style.

▶ Volunteer to draft a portion of a project report, thereby practicing how to think like a client or other stakeholder.

▶ Design a graphic that communicates a complex concept or process, thereby developing an appreciation for how differently individuals respond to visual stimuli.

▶ Listen to the words for facts, and between the words for feelings, before responding; begin to "hear" what is not said.

▶ Examine one of your recently completed projects, identify a portion that would be of interest to your peers, and propose a paper for presentation at a state, regional, or national conference, thereby strengthening your speaking skills and reputation and that of your organization.

Avoid euphemisms, which may amuse but also confuse (Loeffelbein, 1992)

What We Say	*What We Probably Mean*
For your information	I don't know what to do with this, so you keep it
Program	Any assignment that can't be completed in one day
Reliable source	The taxi driver who brought me to work
Give us the benefit of your thinking	We'll listen as long as it doesn't interfere with what we've already decided to do
It is in process	It's so wrapped up in red tape the situation is probably hopeless

Recognize differences between the way women and men communicate in the workplace (Tannen, 1997)

Women Tend To	*Men Tend To*
Use disclaimers: "This may be a silly question, but . . ."	Be assertive: "It is obvious that . . ."; "Note that . . ."
Foresee a crisis, head it off, don't make it a big deal.	Allow a crisis to happen, solve it, and take the credit for the good deed.
Be indirect: "The bookkeeper needs help. What would you think of helping her out?"	Be direct: "Please help the book-keeper."
Give way to a male's interruption, let him take over.	Interrupt, then talk over the person he interrupted.

Listen with eye contact, frequently nodding, indicating understanding and approval.	Listen with stoic expressions, eyes off in another direction, revealing nothing.
Pause, waiting for indications of approval or understanding, and allowing others to contribute.	Speak in a steady stream without pauses. Not allow others to join the conversation.
Solicit opinions before stating a position, to make others feel involved. Say "we."	State viewpoint unequivocally, and take on challengers. Say "I."

Determine the possible influence of body language on some of your recent positive and negative communication experiences

If the words were "right on" but the desired effect did not occur, your body language may not have been aligned with the intended message (Hendricks, 1995; National Institute of Business Management, 1998; Reinhold, 1997).

> *Communication is not what is intended, but what is received by others.*
>
> Mel Hensey

▶ A combination of smiling, relaxed posture, and unrestrained movement suggests happiness or satisfaction.

▶ Frowning, tense posture, and a rigid body or nervous movement indicates unhappiness or dissatisfaction.

▶ Nodding, winking, smiling, and relaxation usually suggest agreement.

▶ Side-to-side head movement combined with frowning and crossed arms project disagreement.

Recognize, both as a receiver and an initiator, the four principal types of communication flow in all types of organizations and what each indicates (Abbott, 1999)

▶ *Downward*, or enabling, communication moves instructions and other directive information down or through a hierarchy.

> *Mental telepathy is not, I fear, a reliable means of communicating in most organizations.*
>
> Charles Handy

▶ *Upward*, or compliance, communication provides feedback to the people who originate downward communication.

▶ *Lateral*, or coordinating, communication moves between peers to maintain or improve operational efficiency.

▶ The *grapevine* fills in gaps in official communication and provides answers to unaddressed questions.

Read the following related lessons

▶ Lesson 11, "Job Security Is an Oxymoron, Career Security Doesn't Have to Be"

▶ Lessons 15 through 24 (Part 2)

Study one or more of the following sources cited in this lesson

▶ Abbott, R.F. 1999. "Downward communication: enabling communication" (Sep. 1); "Upward communication: compliance" (Sep. 8); "Lateral communication: coordination" (Sep. 22); "The grapevine: defying the rules" (Sep. 29). *Abbott's Communication Letter.*

▶ Covey, S.R. 1990. *The 7 habits of highly effective people: restoring the character ethic.* New York: Simon & Schuster.

▶ Hendricks, M. 1995. "More than words." *Entrepreneur* (Aug.), pp. 54-57.

▶ Loeffelbein, B. 1992. "Euphemisms at work." *The Rotarian* (Feb.), pp. 22-23.

▶ National Institute of Business Management. 1988. *Body language for business success.* New York: NIBM.

▶ Reinhold, B.B. 1997. "Body language." *US Airways Magazine* (March), pp. 8-13.

▶ Tannen, D. 1997. *Talking from 9 to 5: how women's and men's conversational styles affect who gets heard, who gets credit and what gets done at work* (audio cassette). New York: Simon & Schuster Audio.

Study one or more of the following supplemental sources

▶ Benton, D.A. 1992. *Lions don't need to roar: using the leadership power of professional presence to stand out, fit in and move ahead.* New York: Warner Books.
    Offers tips on speaking, listening, and visual messages.

▶ Cialdini, R.B. 2001. "The science of persuasion." *Scientific American* (Feb.), pp. 76-81.
    Contends that "six basic tendencies of human response come into play in generating a positive response" to a request. They are reciprocation, consistency, social validation, liking, authority, and scarcity.

▶ Hirsch, H.L. 2003. *Essential communication strategies for scientists, engineers, and technology professionals,* 2nd Ed. Piscataway, N.J.: IEEE Press.
    Stresses the need, whether writing or speaking, to connect to audience needs or wants, establish the flow to take the audience from what they need or want to what we have, and provide reinforcement (that is, substance). Explains the importance of communication knowledge and skill by contending that "communication is the tool by which we operate our socioeconomic system, which is capitalism." Offers some thought-provoking debatable suggestions such as developing a "healthy indifference" to the audience during a presentation and taking a negative position on brainstorming sessions.

▶ Walesh, S.G. 2000. *Engineering your future: the non-technical side of professional practice in engineering and other technical fields,* 2nd Ed. Reston, Va.: ASCE Press.
    Chapter 3, "Communicating to Make Things Happen," discusses listening, three distinctions between writing and speaking, writing, speaking, and body language.

Subscribe to one or more of these e-newsletters

▶ *Abbott's Communication Letter*, a free monthly e-newsletter produced by Robert F. Abbott. Included are practical ideas, presented in an anecdotal format, you can use every day to help you and your organization succeed. To subscribe, go to http://www.abbottletter.com/.

▶ "Qmail," a free questioning-oriented e-newsletter available from Organization Technologies, Inc. To subscribe, contact dleeds@dorothyleeds.com.

In order to speak short on any subject, think long.

Hugh Henry Brackenridge

So, What Do You Know about Bluebirds?

I had six honest serving men—they taught me all I knew:
their names were Where and What and When—and Why and How and Who.
Rudyard Kipling

When our daughter was in high school, my wife and I went with her to her school's annual recognition banquet. Upon arriving, our daughter introduced us to a friend who was there with her parents, and the six of us sat down for dinner.

Initially, there was little conversation. To break the awkward silence, I abruptly asked the group, "So, what do you know about bluebirds?" This question, which came out of the blue, simultaneously embarrassed my daughter and drew some smiles. It broke the ice, though, and we enjoyed a pleasant conversation. (My motivation for asking the question really was to learn about bluebirds, or more specifically, bluebird houses. We were in the process of placing one on our property.)

My daughter frequently kids me about the "embarrassing bluebird question" of years ago. Recalling of the incident, along with considerable subsequent experience, has taught me the benefits of asking questions. One benefit is that it enables mildly introverted individuals (like me) to engage people in conversation in social settings. Most people will answer just about any positive question you ask them and, quite frankly, they seem to enjoy the attention.

In a conversation, keep in mind that you're more interested in what you have to say than anyone else.
Andy Rooney

Asking questions has also helped me throughout my career. On the surface, consulting success might seem to be a function of what you know. In actual practice, however, success is closely linked with how effectively you learn what you need to know. Much of the knowledge you need for a project is acquired by asking many questions about the requirements of the client and the client's stakeholders. Some questions uncover facts and others reveal feelings.

A client once asked me to spend two consecutive days helping a struggling, recently appointed department head. Like many engineers recently promoted to management, this manager was having difficulty with the new responsibility of bringing in work for his department (marketing). After spending a few hours with the new department head, I concluded that he was technically capable, enthusiastic, and a hard worker.

As we continued to talk during that first day, he told me he was scheduled to meet with a potential municipal client the next day to discuss a request for proposal, and he asked if I would accompany him. When I said yes, he enthusiastically grabbed pen and paper and suggested that we prepare for the meeting by listing all the things we wanted the potential client to know about the consulting firm's services.

> *Asking the right questions takes as much skill as giving the right answers.*
> Robert Half

At the risk of dampening his enthusiasm, I offered a counter-suggestion: let's list all the things we would like to know about the client, the client's environment, and the particular project. Furthermore, let's set the goal of having the client talk at least 75% of the time, regardless of how long our meeting lasts. Deferring to gray hair, he agreed.

Our meeting with the potential client lasted one and one-half hours, the client talked at least three-fourths of the time, and we learned much about the potential client, the client's environment, and the project. The firm subsequently got the job. The new manager went on to become a very successful marketer, partly because he switched from a telling mode to an asking mode in his interaction with clients.

As engineers, the quality of our service is a function of what we know about those we serve, their environment, and their projects.

> *My greatest strength as a consultant is to be ignorant and ask a few questions.*
> Peter Drucker

We gain that knowledge by asking thoughtful, open-ended questions that are intended to stimulate informative answers. More bluntly, we don't learn much when we do all the talking. Maybe that's why we have one mouth, two ears, and two eyes. Asking, "So, what do you know about bluebirds?" suggests a powerful means of building personal and business relations.

Suggestions for Applying Ideas

Consider the seven powers of questions (Leeds, 2000)

1. Questions demand answers.
2. Questions stimulate thinking.
3. Questions reveal valuable data and information.
4. Questions put the questioner in control.

5. Questions encourage people to open up.

6. Questions lead to quality listening.

7. Questions cause people to persuade themselves.

Improve your questioning ability by applying some of the following 10 tips

▶ Arrange to meet and ask questions of the right people. Try to talk with knowledgeable people who have decision-making authority.

▶ Prepare for a business conversation by listing questions you would like to ask.

▶ Strive to achieve the 20-80 rule: make statements and ask questions no more than 20% of the time, and listen at least 80% of the time.

▶ Minimize focus on you. Limit use of "I," "me," "we," "our," etc.

▶ Maximize focus on them. Generously use "you," "your," etc.

Children ask better questions than adults.
Fran Lebowitz

▶ Mix close-ended and open-ended questions. Close-ended questions are typically answered with yes, no, or a statement of fact. In contrast, open-ended questions lead to thinking, expressing of opinion, and discussion.

▶ Start open-ended questions with active verbs like explain, clarify, analyze, and translate.

▶ Strive for empathetic listening, that is, acquire information and understand feelings.

▶ Listen for emphasis and respond accordingly. Note the four varied meanings of the following sentence based on which words are emphasized:
  • *I* don't think, at this time, we would be interested in retaining your firm.
  • I don't think, *at this time*, we would be interested in retaining your firm.
  • I don't think, at this time, *we* would be interested in retaining your firm.
  • I don't think, at this time, we would be interested in *your firm*.

▶ Ask questions of strangers.

Recognize the five levels of attentiveness (listed from most to least attentive) (Covey, 1990; Decker, 1992)

▶ Empathetic listening: hearing everything plus understanding feelings (most attentive).

▶ Attentive listening: hearing everything.

▶ Selective listening: hearing only what we want to hear.

▶ Pretending to listen: hearing nothing but looking like we do.

▶ Ignoring: hearing nothing and looking like we hear nothing.

Verify your understanding of what is being said

- ▶ Occasionally say, "I see," "tell me more!" etc.
- ▶ Maintain eye contact.
- ▶ Do not interrupt.
- ▶ Exhibit positive, receptive body language.
- ▶ Paraphrase.
- ▶ Sketch or draw.
- ▶ Ask additional clarifying questions.

Read the following related lessons

- ▶ Lesson 16, "Talk to Strangers"
- ▶ Lesson 46, "Interviewing So Who You Get Is Who You Want"

Study one or more of the following sources cited in this lesson

- ▶ Covey, S.R. 1990. *The 7 habits of highly effective people: restoring the character ethic.* New York: Simon & Schuster.
- ▶ Decker, B., and J. Denney. 1992. *You've got to be believed to be heard.* New York: St. Martin's Press.
- ▶ Leeds, D. 2000. *The 7 powers of questions: secrets to successful communication in life and at work.* New York: Perigee.

Nature, which gave us two eyes to see and two ears to hear, has given us but one tongue to speak.
Jonathan Swift

Refer to the following supplemental source

- ▶ Hensey, M. 1999. *Personal success strategies: developing your potential.* Reston, Va.: ASCE Press.

  Chapter 12 presents a "what I know chart" that can be used to stimulate question-asking. The four levels of knowing outlined in the chart are (1) know that I know; (2) know that I don't know; (3) think I know, but don't; and (4) don't know that I don't know.

Subscribe to this e-newsletter

- ▶ "Qmail," a free questioning-oriented e-newsletter available from Organization Technologies, Inc. To subscribe, contact dleeds@dorothyleeds.com.

Let every man be quick to hear, slow to speak, slow to anger.
James 1:19, The Bible, RSV

Talk to Strangers

You miss 100 percent of the shots you never take.

Wayne Gretzky

*T*he advice suggested by the title of this lesson contradicts what our parents told us to do when we were children. However, we are no longer children and presumably we can take care of ourselves. In her book, *How to Work a Room*, consultant Susan RoAne (1988) challenges readers to "work the world." She urges us to adopt the philosophy that we are surrounded by opportunities to learn and make contacts. But we often have to take the initiative, whether we are at our place of work, doing personal errands in our community, sitting in an airport between flights, or attending a conference.

Will talking to strangers always yield useful information or a new contact? Certainly not. However, RoAne says:

> That's not the point. The point is to extend yourself to people, be open to whatever comes your way, and have a good time in the process. One never knows. . . . The rewards go to the risk-takers, those who are willing to put their egos on the line and reach out—to other people and to a richer and fuller life for themselves.

I tend toward introversion, as do engineers in general, as shown by personality profile studies (Wankat and Oreovicz [1993] and Johnson and Singh [1998]). Nevertheless, I've made the effort to approach strangers in a variety of business and professional settings. Writing this lesson caused me to contemplate some positive results of conversations I've had with strangers. For example, I:

- found out about an effective personal time management system as a result of talking to a stranger on an airplane. I learned that he conducted time management seminars, and he gave me a sample of the system he used.

- obtained a rewarding, multi-year consulting arrangement as a result of approaching a speaker after his presentation at a conference. I complimented him on his presentation, and mentioned that we seemed to have some common acquaintances and shared technical interests.
- received an appointment to a high-level professional society committee partly as a result of introducing myself to a society staff member and suggesting that we share conversation and a cab ride to the airport.

Each lifetime is the pieces of a jigsaw puzzle. For some, there are more pieces. For others, the puzzle is more difficult to assemble. But know this: You do not have within yourself all the pieces to your puzzle. Everyone carries with them at least one and probably many pieces to someone else's puzzle.

Lawrence Kushner

Talking to strangers can be a challenge. It might be difficult to know what to talk about. Frankly, most people like to talk about themselves and their interests. Therefore, give them an opportunity to do so by asking questions. If they are half as thoughtful as you are, they will reciprocate, and many pleasant and informative conversations will result. Even if they don't reciprocate, you'll probably learn something by asking sincere questions. The person who asks questions directs the conversation and benefits the most from it.

If you enjoy a conversation you had with a person you've met, that's a good reason to take action. Exchange business cards. A thoughtful follow-up might be as simple as an e-mail or handwritten note to them saying, "It was a pleasure meeting you. I enjoyed our conversation. I hope to see you in the future."

Suggestions for Applying Ideas

Prepare yourself to start a conversation the next time you enter a reception, icebreaker, or similar event (Staples.com, 2003)

▶ Adopt this mindset: I am here to meet, learn, help, and possibly start mutually beneficial relationships, not sell.

▶ Wear your nametag.

▶ Have a supply of quality business cards handy.

▶ Seek out strangers, not only but mostly.

▶ Prepare a 10-second description of what you do, sometimes called an "elevator speech." Mine is: "I help technical organizations with non-technical problems and opportunities."

▶ Keep moving. Five to 10 minutes per person or small group will work.

▶ Ask for a business card as you meet or are about to leave someone. If they asked you for something or you made a promise, note it on their card.

▶ Listen 80% of the time; talk and ask 20%. As soon as possible after the event, fulfill all promises and make other follow-ups as appropriate.

Ask open-ended questions (i.e., questions that cannot be answered "yes" or "no") when you want to encourage conversation but can't think of a way to start

▶ "What do you hope to get out of this meeting?"

▶ "I notice that you are with Noitall Consulting. What attracted you to the consulting business?"

▶ "I am working on a knowledge management project and need some ideas. Does anybody in your organization do this?"

▶ "I have never been to your part of the country. What is it like to live there?"

Consider these thoughts on serendipity (Bach, 1970) when you hesitate to start a conversation

▶ Serendipity may be defined as the process by which we "dip" into life with "serenity."

▶ Serendipity means "catching on to the magic of the thing called chance," of "making discoveries by accident . . . of things not in quest of."

▶ While we may have a particular goal in mind in talking to strangers, we may serendipitously achieve some other benefit.

Record the essentials of what you learn as strangers become contacts, and sometimes colleagues or even friends

▶ Even if you don't aggressively follow the "talk to strangers" advice of this lesson, you will, in the course of your work and other activities, meet hundreds if not thousands of individuals.

> *You can make more friends in two months by becoming really interested in other people, than you can in two years by trying to get other people interested in you.*
>
> Dale Carnegie

▶ Experience clearly indicates that you will have need to contact many of these individuals for various reasons such as data, information, referrals, employment, and marketing. A contact could occur months, or even years, after meeting a person.

▶ Useful data and information to note about a contact include name, employer, title and credentials, address, telephone and fax numbers, e-mail address, special interests, and names of spouses, children, and mutual acquaintances.

▶ Use some form of electronic system, such as your e-mail directory, to record data and information about members of your growing network.

Read the following related lessons

▶ Lesson 5, "DWYSYWD"

▶ Lesson 9, "Go Out on a Limb"

▶ Lesson 15, "So, What Do You Know about Bluebirds?"

Study one or more of the following sources cited in this lesson

▶ Bach, M. 1970. *The world of serendipity.* Marina Del Rey, Calif.: DeVorss & Company.

▶ Johnson, H.M., and A. Singh. 1998. "The personality of civil engineers." *Journal of Management in Engineering* (July/Aug.), pp. 45-56.

▶ RoAne, S. 1988. *How to work a room: a guide to successfully managing the mingling.* New York: Shapolsky Publishers.

▶ Staples.com. 2003. "Power-schmoozing your way to the top" (Jan. 11), (http://www003.staples.com/content/Entrepreneur/PowerSchmoozing.asp?&).

▶ Wankat, P.C., and F.S. Oreovicz. 1993. Psychological type and learning. Ch. 13 in *Teaching engineering.* New York: McGraw-Hill.

Refer to one or more of the following supplemental sources

▶ Benton, D.A. 1992. *Lions don't need to roar: using the leadership power of professional presence to stand out, fit in, and move ahead.* New York: Warner Books.
    Stresses the need to match professional competence with the ability to make personal connections. Offers advice on physical appearance, greeting strangers, gestures, listening, and voice variation.

▶ Leeds, D. 2000. *The 7 powers of questions: secrets to successful communication in life and at work.* New York: Perigee.
    Claims that questions demand answers, stimulate thinking, reveal information, control the conversation, encourage others to speak, enhance listening, and persuade.

Subscribe to this e-newsletter

▶ "Qmail," a free questioning-oriented e-newsletter available from Organizational Technologies, Inc. To subscribe, contact info@dorothyleeds.com.

No matter how ambitious, capable, clear-thinking, competent, decisive, dependable, educated, energetic, responsible, serious, shrewd, sophisticated, wise, and witty you are, if you don't relate well to other people, you won't make it. No matter how professionally competent, financially adept, and physically solid you are, without an understanding of human nature, a genuine interest in the people around you, and the ability to establish personal bonds with them, you are severely limited in what you can achieve.

Debra A. Benton

You're Tall and That's All

> *Chances are the worst pains you have suffered in life*
> *come from words used cruelly.*
>
> Joseph Telushkin

*I*was the tall, clumsy, uncertain, last player on the high school basketball team. Early in the season, one of my skillful, confident teammates who was in the starting five became frustrated by an example of my ineptitude with the game. He said, "Walesh, you're tall and that's all." Those words hurt badly at the time, partly because they were said within earshot of others; even today, 45 years later, as I write this, I still feel a slight sting. The point: Being called negative names and being the target of mean-spirited or thoughtless words does hurt many of us.

If we are to realize our management and leadership potential and help others do the same, we need to choose our words carefully. Speaking and writing with appropriate sensitivity are complicated by the increase in the complexity and competitiveness of business and other endeavors.

Engineers, especially younger ones, may be especially prone to insensitive use of words. Engineers deal with quantitative factors such as theories, equations, algorithms, and simulation models that determine the outcome of their work. Our communication often is blunt, terse, obtuse, or emphatic, making us seem more insensitive than we intend. But as essential as these objective and quantitative factors are, the success of our efforts is significantly determined by the manner in which we interact with others. Words and the way we use them are at the heart of that interaction process.

> *Man does not live by words alone, although he sometimes must eat them.*
>
> Adlai Stevenson

Too often, support personnel get the brunt of verbal insensitivity. There is an unfortunate tendency, especially among the more technically oriented, to "look down on" or devalue support personnel. This perspective is evident in the way support staff are referred to as "boy" or "girl" or called by their first names without specifically inviting us to do so. Particularly disturbing is to see professionals in the workplace being downright impolite and surly with support staff, but polite to and friendly with the rest of the professional staff. The overall impact is very negative, not only to the morale of the person being degraded, but also to the professional image of the rude individual.

An old Jewish teaching compares the tongue to an arrow, rather than to some other weapon, such as a sword (Telushkin, 1996):

> If a man unsheathes his sword to kill his friend, and his friend pleads with him and begs for mercy, the man may be mollified and return the sword to its scabbard. But an arrow, once it is shot, cannot be returned.

While words, once "shot," can hurt, the opposite also is true. Words can help, heal, and sometimes inspire. For example, perhaps my teammate could have offered a tip on how to improve my basketball playing rather than say my height was my only positive attribute.

Although words exist for the most part for the transmission of ideas, there are some which produce such a violent disturbance in our feelings that the role they play in the transmission of ideas is lost in the background.
Albert Einstein

It is not enough to simply avoid hurtful words. Offering helpful words is the caring thing to do; it's one of the secrets of effective managing and leading. Think of the enabling power of words of encouragement that you have received from a parent, a sibling, a friend, a spouse, a teacher, a colleague, or a client. Often, those words—merely vibrations in the air—make all the difference.

The words we use are rarely neutral in their impact on others. Words and the accompanying tone, facial expression, and body language either help or hurt the recipient. Let's remember to put our brains and hearts in gear before engaging our mouths.

Suggestions for Applying Ideas

Ask your spouse, a family member, or a close friend to critique the words you use and how you use them

- ▶ How have your words hurt them?
- ▶ How have your words helped them?
- ▶ How could you more effectively use words?

Read the following related lesson

- ▶ Lesson 45, "Our Most Important Asset"

Study the source cited in this lesson

▶ Telushkin, J. 1996. "Words that hurt, words that heal: how to choose words wisely and well." *Imprimis* (Jan.).

There is one whose rash words are like sword thrusts,
but the tongue of the wise brings healing.
Prov. 12:18, The Bible, *RSV*

18

Balance High Tech and High Touch

*All of the technology in the world will not help us
if we are not able, at the core, to communicate with each other
and build strong, lasting relationships.*

Dorothy Leeds

*T*oday's electronic communication and data-gathering devices are amazing. Many of us frequently, if not daily, use voicemail, teleconferencing, fax, e-mail, pagers, and the World Wide Web. The next wave is already surging: more wireless tools; live, multi-station audio-video conferencing; "smart" documents; and individualized, self-paced, web-based education and training. And who knows what is beyond that.

We derive numerous and varied business and professional benefits, such as saving time and money, from this electronic "wizardry." For example, we can

- Function as virtual teams and departments within and among our businesses, even though we work in widely scattered locations.
- Work on national committees within our professional organizations.
- Conduct research using global resources.
- Manage multi-office projects and programs.
- Recruit personnel and market services.

There is a potential downside, however, to over-reliance on electronic communication technology: It can actually reduce interpersonal communication. We can lose meaningful contact with our stakeholders, that is, co-workers, business partners, clients, regulators, and citizens.

Engineers and other technical personnel are particularly susceptible to overuse of electronic communication devices because of our tendency to introversion (Johnson and Singh, 1998; Wankat and Oreovicz, 1993). Excessive use of electronic devices can push us even closer to the introvert end of the introvert-extrovert spectrum and, as a result, take us further from meaningful, empathetic (facts plus feelings) interaction with others. As technical professionals, we may be tempted to communicate electronically, rather than personally, whenever we can. After all, if some electronic communication is good, isn't more even better?

> *Technology does not improve the quality of life; it improves the quality of things. Improving the quality of life requires the application of wisdom.*
>
> *Neil Armstrong*

Success in engineering and other technical professions is ultimately based on the trust we earn from our stakeholders. Earning and maintaining trust requires some meaningful human interaction, that is, "face time," "eyeball-to-eyeball" conversation, and empathetic listening. Communication that is efficient and uses the latest electronic devices is not necessarily the most effective, especially when it becomes the only mode of interaction.

Electronic devices are not communication. They are tools used in communication. Let's frequently and skillfully use electronic tools but intermix them with human interaction. Let's balance their high tech with our high touch.

Suggestions for Applying Ideas

Personalize the communication between yourself and others

▶ Before establishing another committee, project team, or virtual department within your multi-office organization, get the members together for a working and social event. Create an opportunity for them to become acquainted and to connect. Then expect them to communicate primarily by electronic means.

▶ As you begin to compose yet another e-mail to a client or someone you serve, pick up the telephone instead and suggest that the two of you "do lunch."

▶ Rather than send a generic thank you e-mail or memorandum to the members of your team for a job well done, compose a personalized, handwritten note to each person and send it via "snail mail." Better yet, hand it to them with a brief statement of appreciation

Read the following related lessons

▶ Lesson 14, "Communicating Five Ways"

▶ Lesson 17, "You're Tall and That's All"

▶ Lesson 34, "TEAM: Together Everyone Achieves More"

▶ Lesson 35, "Virtual Teams"

Study the sources cited in this lesson

▶ Johnson, H.M., and A. Singh. 1998. "The personality of civil engineers." *Journal of Management in Engineering* (July/Aug.), pp. 45-56.

▶ Wankat, P.C., and F.S. Oreovicz. 1993. Psychological Type and Learning. Ch. 13 in *Teaching engineering*. New York: McGraw-Hill.

The more people are reached by mass communication,
the less they communicate with each other.

Marya Mannes

19

Trimming Our Hedges

> *On speaking, first have something to say, second say it,*
> *third stop when you have said it.*
> John Shaw Billings

Engineers are, by and large, decent individuals. They want to be open and honest in their interactions with others. That desire, taken to extremes, leads to overexplanation of results that others may perceive as a lack of competence, confidence, or commitment.

What is needed is some "hedge trimming": the ability to speak, write, and answer questions in a positive manner, to present our views without excessive qualifications, so the information is presented at the level appropriate for our intended audience.

For example, at a staff meeting, you would not answer a question about the required size of an electric motor by saying, "If I did the calculations correctly ..." You are responsible to do the calculations correctly, or to find out how to do them. In contrast, a qualified answer—"Based on the limited field data, in my opinion there will be no foundation problems"—is an acceptable response to questions from professional peers who understand the complexity of our work.

As another example, that same statement ("Based on limited field data ...") would not be appropriate to present to a public body because it may be beyond the area of expertise of the audience. The "in my opinion" part of the statement is sufficient.

> *Mastery of language affords remarkable power.*
> *Frantz Fanon*

What we say and how we say it influences others' opinions of us, our work, and the organizations we represent. The tendency to overqualify statements and responses suggests the following negatives to listeners: inadequate preparation, lack of ability, low self-confidence, or insensitivity to colleagues and clients. These perceived traits can detract significantly from the image and performance of the technical professional. Unrectified, this could interfere with advancement within the organization and the profession. More importantly, these negative impressions can thwart the implementation of viable recommendations. The fact that we are well prepared, have ability, and are confident and competent is irrelevant if we are perceived to be otherwise. Perception is fact.

We should do all assignments well. When explaining or reporting the results to others, we must be sensitive to the nature and interest of the audience. Let's speak in simple, declarative, and brief fashion, unencumbered with excessive, convoluted qualifications. Less hedging leads to more personal commitment. Let's trim our hedges.

> *The ancestor of every action is a thought.*
> Ralph Waldo Emerson

Incidentally, the preceding focuses on how what we say influences others. What we say and how we say it also influences ourselves. Consider the more positive effect on your subsequent performance when you say, "I will get the draft report to you by Friday noon" rather than "I will try to get the report to you by Friday noon." Less hedging leads to more personal commitment. Let's trim our hedges.

Suggestions for Applying Ideas

Listen for hedges in your communication

Table 19-1 illustrates how the words we use in routine conversation set high or low expectations for ourselves and those around us (Hill, 1960; Rogers and McWillams, 1991; and Walther, 1991).

Study the following sources cited in this lesson

Hill, N. 1960. *Think and grow rich*. New York: Fawcett Crest.

Roger, J., and P. McWilliams. 1991. *Do it! let's get off our butts*. Los Angeles, Calif.: Prelude Press.

Walther, G. R. 1991. *Power talking: 50 ways to say what you mean and get what you want*. New York: Berkley Books.

> *For one word a man is often deemed to be wise,*
> *and for one word he is often deemed to be foolish.*
> *We should be very careful indeed what we say.*
> *Confucius*

Table 19-1 *How Words Set Expectations*

Indicates Low Expectations	Indicates High Expectations
I'll have to ...	I'll be glad to ...
I'll try to ...	I will ...
I can't do ...	I haven't done ... but want to learn how
This is a problem	This is an opportunity
I will spend on education and training	I will invest in education and training
This is impossible	This is going to require a special effort
Excuse my messy office	Welcome to my place
I've just been here one month but maybe...	I've observed the situation since starting a month ago and suggest ...
I got lucky	I set goals and acted constructively to achieve them
I performed poorly	I learned ...
If only I ...	I will ...
She just doesn't understand ...	I will try another way to explain ...
That's your problem ...	Let's try ...
This should take a few weeks ...	I will get it to you by ...
To tell the truth ...	(Just tell it)
You don't remember me ...	Good to see you again, my name is ...

Write to Find Out What We Think

> *I write to find out what I'm thinking.*
> Edward Albee, playwright

One stereotype about engineers is that they are poor writers. In my experience, however, as a project manager, coach, consultant, and engineering professor, engineers can become effective writers, like most reasonably intelligent and motivated individuals. The "secret" is an old but powerful formula: high expectations and high support.

For several years, I incorporated formal writing assignments in an introductory transportation engineering course to sophomore civil engineering students. Each student was required to research a transportation structure, facility, or system of their choice and write a report addressing the topics of milestones, technical challenges, legal requirements, environmental issues, safety provisions, financing, proponents and opponents, and lessons learned. The assignment required a product that was professional in appearance and content (high expectations). Numerous but brief writing tips and aids were provided as the semester proceeded, and a sample chapter was carefully critiqued (high support). Most of the reports received A's and B's even though I was a very demanding grader. In many cases, the content and appearance of the student reports exceeded that of reports written by professional engineering firms and government entities.

Managing and leading effectiveness is greatly enhanced by writing competency. The writing assignments were intended to convey to students the importance of effective written communication for successful and satisfying professional practice. The writing project also was intended to provide students with more researching, organizing, and writing skills while building their confidence as writers.

A third purpose of the assignments was to suggest that writing, in addition to being a means of interpersonal communication, also is one very effective form of *intra*-personal communication. Writing about our concerns, questions, ideas, theories, and options helps to clarify our thinking and illuminate the darker recesses of our minds. It can be powerful preparation for engaging and influencing others. As publisher Malcolm Forbes said, "Putting pen to paper lights more fire than matches ever will."

> *The blank page on which I read my mind.*
> Dylan Thomas

A personal "writing to learn" example I like to share is as follows. Years ago, I became interested in how the world of work might change as we move into the 21st century. What will the engineering work force look like? Who will be served by our profession? What needs will we fulfill? How will work methods change? What won't change? As a means of finding answers to these questions, I committed to presenting a paper on the subject before I had done the research. My subsequent research was successful: I learned much, wrote and presented the paper (Walesh, 1992), and have frequently been able to use what I learned. In fact, that paper later became a section in one of my books (Walesh, 2000).

I suggest using focused writing—that is, selecting a subject of interest and commiting to writing and then speaking about it—to facilitate improved self-understanding and learning, which leads to more effective speaking. Encourage others to do the same.

On the surface, we probably think of writing as one means of communicating with others. And it is, along with the mechanisms of listening, speaking, visuals, and mathematics—the five means available for interpersonal communication. But even more fundamentally, writing at least for some of us, me included, is a powerful means of intra-personal communication, communication with ourselves. Somehow, writing about our concerns questions, ideas, theories, and options helps one "part" of us communicate with another "part" of us. Maybe it's our "brain" communicating with our "heart" or our "intuition" interacting with our "reason." Perhaps writing helps our left or logical brain communicate with our right or intuitive brain. Effective intra-personal communication, achieved through writing, can be a powerful preparation for effective inter-personal communication.

One of the reasons for writing essays like the one you are reading, and it's a selfish one, is to find out what I think—really think—about various non-technical aspects of professional life. Sometimes I am surprised.

We should all do more writing, partly to refine our skills and make us better communicators and partly to clarify our thinking. However, a blank sheet of paper or a blank screen can be foreboding. It strikes great fear in some of us. As we begin to push our pen or tap the keys, our thoughts focus. Previous related experiences, good and bad, are relived. Lessons learned are recalled. Important questions arise. Accepted premises are challenged. Creative juices begin to flow. Write on! Right on!

> *You learn as much by writing as you do by reading.*
> Eric Hoffer

Suggestions for Applying Ideas

Enhance your work with clarification writing, as these two examples illustrate

► As a member of a project team, you encounter a technical problem. You've discussed it with a few colleagues, the problem is ill defined, and you have some vague ideas on how to resolve it. Now is the time to write. Describe the problem, separating symptoms from causes. Your understanding of the problem will deepen as you write about it. Based on your improved, but probably not complete, understanding of the situation, list, cluster, outline, and write alternative solutions, noting some pros and cons of each. Your creativity will flow. Write out your recommendations, at least in a tentative fashion. Obstacles and opportunities will appear. List, cluster, outline, and write about them, in at least a preliminary fashion. Share the resulting draft document with team members. The mental discipline inherent in the writing process will add great value to your team's efforts.

► As a technical professional given the responsibility for marketing some of your firm's services or products to new clients or customers, use writing to discipline your approach. List, cluster, and write what you know about the client's or customer's organization and its personnel. Based on this, indicate what you would like to know. Describe how you will obtain that information. A mini-marketing plan now lies on your desk or appears on your monitor. Great value has been leveraged from the half-hour of writing, or more precisely, thinking. Now carry out the plan, refining it, in writing, as you proceed.

Prevent the editor within you from inhibiting the creator

► Write for a predetermined time, say 10 minutes or one-half hour, without stopping to correct or reword your writing.

► Go first for content; editing can follow.

Consider William Zinsser's idea (1988) that writing is a powerful learning process

► Speaking about the ideas of people representing a wide variety of interests and disciplines and learning about what these individuals do, Zinsser says, "I've found that their ideas were never so specialized that I couldn't grasp them by writing about them or by editing someone else's writing about them."

► About colleges and universities, and their writing across the curriculum programs, Zinsser says, "I saw that 'writing across the curriculum' wasn't just a method of getting students to write who were afraid of writing. It was also a method of getting students to learn who were afraid of learning." He continues, "Writing could get into corners that other teaching tools couldn't reach."

▶ We should encourage more "writing across our public and private engineering organizations" to stimulate our personnel to learn more about what we do and why and how we do it. Spread the meeting minutes, report, proposal, and other writing assignments and watch writing skills improve while knowledge of the organization increases.

▶ Emphasizing the power of writing, Zinsser says, "Writing enables us to find out what we know—and what we don't know—about whatever we're trying to learn."

Read the following related lessons

▶ Lesson 3, "Smart Goals"

▶ Lesson 21, "Start Writing on Day 1"

▶ Lesson 32, "The Power of Our Subconscious"

Study one or more of the following sources cited in this lesson

▶ Walesh, S.G. 1992. "Changing demographics: civil engineering applications." Presented at the 1992 International Convention and Exposition of the American Society of Civil Engineers. New York: ASCE.

▶ Walesh, S.G. 2000. *Engineering your future: the non-technical side of professional practice in engineering and other technical fields,* 2nd Ed. Reston, Va.: ASCE Press. See the section titled "The Changing World of Engineering and Other Technical Work" in Chapter 15.

▶ Zinsser, W. 1988. *Writing to learn.* New York: HarperResource.

Subscribe to this e-newsletter

▶ "A.Word.A.Day," provided free daily by Wordsmith.Org. Presented in this newsletter is a word with its definition and an example of its use. To subscribe, go to http:www.wordsmith.org/awad/.

Writing is thinking on paper.
William Zinsser

Start Writing on Day 1

> *Begin with the end in mind.*
> Stephen Covey

*I*f a project's deliverables include some form of report, start writing the report when the project starts. The writing process should parallel the performance of the work, and draft portions of the report should be provided to the client and other stakeholders as they are produced. Experience suggests three benefits of writing and delivering in parallel to working:

Improved communication within the project team. Writing and reading draft descriptions of project features—such as background, purpose, scope, and approach—are likely to raise useful intra-team concerns, questions, and suggestions. Early drafts prompt clarification of vague aspects of the project.

Improved communication with the client and other interested parties. This applies to areas such as overall report format and content, level of report detail, thoroughness of data and information sources, assumptions made, approaches taken, alternatives being considered, and recommendations considered. The client receives and responds to a steady flow of data, information, assumptions, alternatives, and recommendations rather than being blindsided with voluminous material (some of which is probably off target) near the end of the project.

Improved accuracy and quality of reports and project information. Errors and omissions that result from "last minute" rushing are avoided. Deficiencies perceived from the appearance and content of reports can diminish the perceived quality of the project. Sadly, engineers and other technical professionals often deliver excellent work in mediocre packages and the quality of the effort is judged by the appearance of the package.

> *Some of the most accomplished engineers of all time have paid as much attention to their words as their numbers, to their sentences as to their equations, and to their reports as to their designs.*
>
> Henry Petroski

A common objection to starting the writing on day 1 is that there is nothing to write about because no work has been done. However, a draft report outline can be prepared immediately, during the first 5 to 10% of the project budget or duration. The initial outline becomes the framework on which the report is constructed.

Furthermore, background, purpose, and scope subsections can be drafted immediately. At the very start of a project we should know or understand these items. If we don't, our project is already in trouble. Immediately seek definition and clarification in writing, before proceeding further. A glossary, list of abbreviations, and list of cited references also can be started. If, after the project has passed the 10% mark, there is still nothing to write about, then perhaps nothing useful is being accomplished.

Suggestions for Applying Ideas

Locate documentation guidelines

▶ Determine if either the organization you work for or your client has a style guide or procedural manual for writing reports, memoranda, and other documents.

▶ Some companies publish internal guidelines that specify content for all the reports and documents required for a project. These might even be available in electronic "template" or "boilerplate" format.

▶ Because reports and other documents are often written as a team effort, writing guides help achieve consistency and reduce subsequent editing.

Obtain and use recognized writing guides if your organization does not have one

▶ The following guides typically address topics such as abbreviation, capitalization, citation of references, gender-neutral writing, punctuation, typestyles, graphics, common mistakes, and overall format:
  • *The Chicago Manual of Style (2003)*
  • *The Elements of Style (1979)*

Lead the creation of a writing guide or similar document if your organization does not have one

▶ The return on investment will be great because of the resulting consistency within documents and reduced editing efforts.

▶ Tailor the guide to the needs of your organization, making it long and detailed enough to be useful but not so extensive as to be cumbersome.

▶ Refer to existing, readily available style guides, such as those mentioned previously, prepared by organizations similar to yours, or listed in the sources section of this lesson. With appropriate permission, utilize them as examples.

Share the report outline and style guide with stakeholders

▶ Examples of possible report or document stakeholders are colleagues, sub-contractors, subconsultants, vendors, client or owner representatives, regulators, and elected and appointed officials.

▶ By sharing the report's content, format, and style with key stakeholders, and giving them an opportunity to comment, we are less likely to blindside individuals and entities and more likely to meet requirements.

Design reports and other documents to communicate with the multiple audiences likely to read at least a portion of the product

▶ Provide a brief *Executive Summary or Abstract* near the beginning of the document. Write it after the report is essentially complete, focusing on principal findings and recommendations. Many readers will read only the Executive Summary. Keep it short and write it well.

▶ The *body* of the document, perhaps consisting of chapters or sections, can include topics such as background, purpose, scope, related studies, description of project area/site/problems, evaluation of options/alternatives, costs, benefits, implementation steps, recommendations, acknowledgments, and cited references. Relatively few readers will digest the body of the document. However, those that do will tend to want a comprehensive understanding of the project.

▶ Conclude the document with *appendices/attachments/exhibits*. Include detailed supporting material such as cost estimates, derivation of equations, computer program input/output, and copies of key documents. Few readers will study all of these supporting items; however, those who do will closely examine at least some of them.

Brainstorm, list, cluster, outline, and incubate the next time you manage a report writing effort

▶ Place a blank sheet of paper on your desk and begin to brainstorm your topic. Quickly write, in the form of single or a few words placed anywhere on the paper, whatever comes into your mind. The paper will soon be filled with ideas and information. Put the project aside for a while to engage your subconscious mind and then resume the brainstorming and recording.

▶ Arrange the ideas and information into clusters or groups. This will bring some order to the results of your brainstorming effort.

▶ Use the clusters or groups to outline the report, in ever-increasing detail, in two or more separate sittings. "Two or more" sittings gives your subconscious mind an additional opportunity to incubate the ideas you feed it via the outline.

▶ Resist the temptation, at this early stage, to write complete sentences and paragraphs. Focus, instead, on a crude outline that you revisit at least one or two times.

Writing is a tool that enables people in every discipline to wrestle with facts and ideas. It's a physical activity, unlike reading. Writing requires us to operate some kind of mechanism—pencil, pen, typewriter, word processor—for getting our thoughts on paper.

William Zinsser

▶ Each time you pick up the outline, you are very likely to see the results of your subconscious mind's work—new insights, ideas, and content.

▶ Share the evolving outline with team members and possibly other stakeholders. Give them and their subconscious minds time to work on the outline.

▶ After receiving input from others, refine the outline one more time. Resist being prematurely bound by the outline. It's just words on paper, so freely delete, insert, and rearrange.

▶ Listen to your writing, that is, read your outline out loud. Revise as needed.

▶ Draft the report. This might be a one-person task for review by others or a team effort.

▶ Listen to your writing, that is, read your report out loud. Revise as needed.

Read the following related lessons

▶ Lesson 20, "Write to Find Out What We Think"

▶ Lesson 32, "The Power of Our Subconscious"

Study one or more of the following sources cited in this lesson

▶ Berthouex, P.M. 1996. "Honing the writing skills of engineers." *Journal of Professional Issues in Engineering Education and Practice* (July), pp. 107-110.

▶ Bowler, P.L. 1990. *Readable writing handbook*. Helena, Mont.: Montana Dept. of Highways.

▶ Evans, M.D. 1995. "Student and faculty guide to improved technical writing." *Journal of Professional Issues in Engineering Education and Practice* (April), pp. 114-122.

▶ Neville-Scott, K. no date. *Technical writing/communications workshop: taking the mystery out of writing*. Cambridge, Mass.: Camp Dresser & McKee, Inc.

▶ Strunk, W., Jr., and E.B. White. 1979. *The elements of style,* 3rd Ed. New York: Macmillan.

▶ University of Chicago. 2003. *The Chicago manual of style,* 15th Ed. Chicago, Ill.: University of Chicago Press.

▶ Walters, R., and T.H. Kern. 1991. "How to eschew weasel words and other offenses against logic." *Johns Hopkins Magazine* (Dec.), pp. 25-32.

Refer to one or more of the following supplemental sources

▶ Petroski, H. 1993. Engineers as writers. *American Scientist* (Sept./Oct.), pp. 419-423.

Noting that "according to conventional wisdom, engineers eschew writing, reading, and speaking," this paper contends that engineers spend a significant fraction of their work time "writing, editing, or preparing oral reports." Presumably as a means of motivation, if not inspiration, the author identifies engineers known for their effective writing. Included are Vitruvius, Caesar's chief engineer and architect; mining engineer and later 31st U.S. President Herbert Hoover and his wife Lou Henry Hoover; writer and engineering professor Alexander Low Waddell; electrical engineer Charles Steinmetz; bridge designer Othmar Ammann; Thomas Telford, first president of the British Institution of Civil Engineers; and Joseph Strauss, first proponent and then chief engineer of the Golden Gate Bridge.

▶ Walesh, S.G. 2000. *Engineering your future: the non-technical side of professional practice in engineering and other technical fields,* 2nd Ed. Reston, Va.: ASCE Press.

Chapter 13, "Communicating to Make Things Happen," includes a large section containing report writing suggestions, such as defining audiences, outlining and incubating, writing easy parts first, employing a gender-neutral style, using rhetorical techniques, avoiding tin ear, adopting a flexible format, using lists, establishing report milestones, and citing sources.

Subscribe to this e-newsletter

▶ "Word of the Day" is provided free by Lexico Publishing Group. This daily e-newsletter defines a new word and illustrates it with quotes. To subscribe, go to http://www.dictionary.com/wordoftheday/list/.

▶ "Wordbiz Report" is offered free by WordBiz.com. Offers practical tips for more effective writing. Subscribe by going to http://www.wordbiz.com

I didn't have time to write a short report; that's why this one is so long.

Anonymous

P⁵: Preparing, Presenting, and Publishing Professional Papers

Whenever an engineer learns something new in technics,
it is his bounden duty to put it in writing and see that it is published
where it will reach the eyes of his conferees, and will always
be available to them. It is absolutely a crime for any man to die
possessed of useful knowledge in which nobody shares.

John Alexander Low Waddell

Preparing, presenting, and publishing professional papers (P⁵) is a concrete means of giving something back to the engineering profession. P⁵ also provides the following personal and organizational benefits:

- Improved writing and speaking ability, which is directly and immediately transferable to many aspects of our professional and personal lives.
- Increased confidence as a result of interacting with peers.
- Expanded visibility of our organization with emphasis on its accomplishments and abilities.
- Earned membership in networks of leaders, which provides quick access to assistance when needed.

Presenting and publishing may sound like a great idea with clearly evident benefits—a win-win-win-win situation all around for the audience, the profession, your employer, and you. But you probably think you don't have the time to prepare, present, and publish a paper. In actuality, the extra investment in time and effort to produce a paper typically will be very small compared to the substantial effort you and others have already expended on a potential paper's subject. The following paragraphs describe and thereby simplify the P⁵ process.

Content. You do not need to first do special work that would serve as the subject for a paper. Instead, you and potential co-authors should seek to write papers about the good work that you are doing or already have done. Aspects of current projects, which may seem routine to you because of your immersion in them, may be exciting and valuable to your peers. Scan your current or recently completed projects for potential topics.

Of course, you may be doing work the results of which are not worth sharing with the profession. If all or most of your efforts are producing such marginal results, are you practicing sound stewardship with your education, experience, and potential? If not, perhaps you should lead internal changes or move on to an employer whose work is worth speaking and writing about.

Getting on the Program. With your topic in hand, seek possible conferences for presentation of your as-of-now unwritten paper. Learn about upcoming conferences by noting published calls for papers, reading articles in professional society periodicals, skimming flyers announcing conferences, visiting websites of professional organizations, and using your network.

Having identified one or more appropriate conferences, write and submit an abstract of your still unwritten paper. Theoretically, the abstract should be written after the paper is drafted. Realistically, the abstract is usually written before the paper is written. This backward approach works well provided that you already have, as a result of your project involvement, the material you will need and that you have thought through the structure of the presentation, at least in a conceptual fashion.

Writing the Paper. Assuming your abstract is accepted, immediately begin to outline and write the paper. Do so even if a written paper is not a condition for oral presentation at the conference. The following are benefits of developing a written paper:

- Having exercised the discipline and applied the thought required to prepare a written paper, we are **better prepared** for the oral presentation.
- The written paper can be provided, electronically or in hard copy, before or after the presentation to anyone who requests it or to individuals we think might be interested. This use of the written version *expands the paper's audience and influence,* strengthens our personal network, and enhances the reputation of our organization.
- Having the written paper available facilitates the usually desirable last step in the P^5 process: *publication* in a journal or periodical.

> *The courage to speak must always be balanced by the wisdom to listen.*
> Benjamin Franklin

Presenting the Paper. As the presentation time approaches, don't assume anything. Verify audio-visual equipment. Check out the presentation room at least one hour before your presentation or the start of the session containing your presentation. Prepare a short written biographical sketch that the host or session chair can use to introduce you. Speak directly to the audience, not to the screen, your notes, or the ceiling. Prompt questions, answer those you can, and follow up on the others.

Publishing the Paper. Some conferences produce proceedings, which is a collection of written versions of the presented papers. Having one's paper in the proceedings helps to extend its longevity, expand its availability, and leverage its influence.

If a proceedings is not produced, the organization sponsoring the workshop is likely to publish journals or other periodicals. Determine which of these is most appropriate and obtain the paper submittal requirements. Revise the written paper to comply with those requirements and to reflect input received during and after the oral presentation. If the sponsoring organization does not produce journals or other periodicals, find an appropriate professional organization that does and work with them.

Suggestions for Applying Ideas

Look for potential topics for papers among your current or recently completed projects

► A computer simulation model.

► A new technology or a new application of an existing technology.

► An innovative process to more effectively involve stakeholders in decision-making.

► A procedure for conducting benefit-cost analyses on infrastructure projects.

► An unusual means of funding a research and development project.

Structure your overall presentation around T³

► Tell them what you are going to tell them.

► Tell them.

► Tell them what you told them.

Ask yourself what you would like members of the audience to do as a result of hearing you

► Some possible answers:
  • Try a design approach that you found useful.
  • Take at least one step toward developing more leadership ability.
  • Advocate a change in state engineering licensing laws.
  • Use your firm's services or products.
  • Purchase a particular product.

> *A speech without*
> *a specific purpose is*
> *like a journey without*
> *a destination.*
> Ralph C. Smedley

▶ Align all aspects of your presentation with the desired action or actions of your audience. That is, as you determine what to say and how to say it, consider what you want audience members to do.

▶ Explicitly tell your audience what you hope they will consider doing as a result of your presentation. This should be an integral part of the T^3 mentioned previously.

Recognize and prepare for the different preferred learning and understanding styles that are likely to be present in your audience

▶ Basic learning styles are:
- Auditory: Understand mainly by hearing.
- Visual: Understand principally by seeing.
- Kinesthetic: Understand mainly by touching and doing.

▶ Prepare for the visually oriented members of your audience by using photographs, line drawings, graphs, cartoons, icons, and videos.

▶ Interact more effectively with the visual learners by selecting colors to symbolize or convey messages. Some possible examples are listed below (Raskin, 2002):

- Blue   Trust, authority, security
- Green   Money, growth, environment
- Orange   Movement, construction, energy
- Pink   Femininity, calm
- Purple   Royalty, spirituality
- Red   Power, activity, rescue
- Yellow   Light, future, philosophy

▶ Communicate with the visual- and kinesthetic-oriented participants by using props, such as an example of construction material, a piece of equipment, a model of a structure or facility, or a software demonstration.

Leverage your published paper or article

▶ Send copies to individuals who participated in or contributed to the work in the paper, thanking them for their help.

▶ Insert copies, as appropriate, in your firm's statement of qualifications, proposals, and other marketing materials.

▶ Use your organization's website or newsletter to advise clients and others of the availability of the paper.

▶ Prepare variations on the paper, possibly including new developments or emphasizing other aspects, and use the revised paper as the basis for a new presentation-publication cycle.

▶ A word of caution: To avoid violating any copyright agreements you might have signed with the proceedings publisher, obtain permission to reproduce or reprint as-published materials.

Read the following related lessons

▶ Lesson 23, "Practice Out Loud"

▶ Lesson 49, "Giving to Our Profession and Our Community"

Study the source cited in this lesson

▶ Raskin, A. 2002. "The color of cool." *Business 2.0* (Nov.), pp. 49-52.

Refer to one or more of the following supplemental resources

▶ Advanced Public Speaking Institute. 2003. "Public speaking: stage fright." *Leadership and Management in Engineering* (Jan.), pp. 4-5.

Recognizes stage fright, noting that some fear is helpful. Offers 16 visualization tactics that can be used anytime and 36 tactics for possible use just before you speak.

▶ Carnegie, D. 1935. *Public speaking and influencing men in business.* New York: Association Press.

Although old, as is evident from use of the word "men" in the title, this book offers many ideas to help you prepare for and deliver a speech. The underlying message is that there is no "silver bullet" for becoming an effective speaker; adopt sound principles and practice, practice, practice.

▶ Creed, M.W. 1999. "Maximum impact: organizing your presentation." *Journal of Management in Engineering* (Sep./Oct.), pp. 28-31.

Describes three models that can be used when organizing a presentation. The models help the speaker define what is to be accomplished and how.

▶ Walesh, S.G. 2000. *Engineering your future: the non-technical side of the professional practice of engineering and other technical fields,* 2nd Ed. Reston, Va.: ASCE Press.

Chapter 3, "Communicating to Make Things Happen," offers practical tips on how to prepare for and deliver an effective presentation. Topics include defining the audience and the setting, preparing the script, creating graphics, practicing out loud, arranging for and verifying audio-visual equipment, suggesting a proper introduction, delivering the speech, prompting questions and comments, and following up.

I love being a writer. What I can't stand is the paperwork.

Peter De Vries

Practice Out Loud

Practice is the best of all instructors.
Publilius Syrus

*E*ngineers have numerous opportunities to make or help to make presentations. Typical audiences are colleagues within our organizations, clients and potential clients, members of professional associations, service clubs, environmental groups, and students. The professional and business community places a high premium on the ability to make effective presentations. Individuals who develop good to excellent speaking skills are well regarded as measured by span of influence, promotions, compensation, perquisites, added opportunities, and, perhaps most important of all, personal satisfaction for a difficult task well done.

One of the most powerful ways to prepare for delivering an effective presentation is to practice it "out loud." Practicing out loud means exactly that, not "saying" the words silently to yourself.

Practice your presentation out loud, including use of visuals and props, preferably in front of one or more people with whom you feel comfortable and who are likely to offer constructive criticism. If you are not able or inclined to do that, practice your presentation out loud in the privacy of your office, home, hotel room, the room in which you will make the presentation, or some other location. Never give a presentation without having practiced it out loud several times, even if only out loud to yourself. Equally important, never memorize a presentation. An audience will immediately detect memorization and will cease to listen attentively.

There are three reasons to practice a presentation out loud:

1. Out loud practice **establishes the actual time** required to give a presentation. Reading or speaking the presentation to ourselves makes us likely to greatly underestimate the delivery time. Most presenters are given a specific

> *If I am to speak ten minutes, I will need a week's preparation. If 15 minutes, three days. If half an hour, two days. If an hour, I am ready now.*
>
> Woodrow Wilson

time allotment. If you underestimate the time required to deliver your talk, you risk going over and, in turn, agitating the audience, offending other speakers, and embarrassing yourself and your organization.

2.   Each time we practice a presentation out loud, we *become more knowledgeable* about the content of our speech. We discover additional words and phrases to better say what we want to say. We master words that were initially difficult to pronounce. We discover aspects of our topic that require further investigation or elaboration. When we actually make our presentation to the audience, we will, accordingly, have more experience to draw on. We will already have given a similar speech several times. Many good speeches that have been well prepared in terms of content and organization are spoiled because the speaker seems to be searching for the right words. Although this probably reflects the lack of out loud practice, not a lack of substantive content or familiarity with material, the audience is likely to perceive the latter.

3. Practicing a presentation out loud to trusted colleagues or to family members helps us *identify and reduce distracting habits*. Playing with change or keys in our pockets, interspersing hedges such as "ah" and

> *How can I tell what I think till I see what I say?*
>
> E. M. Forster

"you know," avoiding eye contact, looking at just a portion of the audience, and speaking to the screen instead of the audience—all are examples of distractions. Others are mispronouncing words, speaking in a monotone, using unusually long sentences, rocking back and forth, and frequently taking off and putting on glasses.

Another way to practice is to speak frequently on a variety of topics in a variety of settings to a variety of audiences. Each time we speak, receive critique and audience feedback, and act on the results, we become more effective speakers.

Good to great speakers, like good to great actors, athletes, and musicians, make it look easy, and we as listeners and viewers benefit. Experience suggests, however, that what looks easy is the result of hard work, part of which is practicing out loud.

Suggestions for Applying Ideas

Make an audio recording, or better yet, an audio-video recording, of your practice session

► This is especially useful if you are reluctant or unable to practice your presentation out loud with a live audience.

▶ A simple way to record the audio for at least a portion of your presentation is to call your home or office telephone number and leave a voicemail that is all or a portion of your out loud practice. What you hear may not sound the way you think it does.

▶ A video recording offers the added benefit of enabling you to see how you say what you say. Body language is an essential element. Assume, for example, that you inadvertently and often cross your arms across your chest while speaking. Many audience members will interpret this as a signal that your mind is made up and that you are not open to other perspectives or options, even though your words say otherwise. What you say and how you appear when you say it should be aligned.

Experiment with techniques you admire in accomplished speakers

▶ Technique examples include gestures, props, alliteration, unusual visuals, varying volume and pace, and moving into and out of the audience.

▶ Study effective speakers in person, on television, or on videotapes. Note the techniques that work for them.

▶ View each of your presentations as an opportunity to experiment with at least one new approach. Try mimicking some of the techniques you've seen and heard other speakers use, with suitable modifications for your personality and the particular presentation.

Experience the real sound of your voice

▶ Unless you have listened to and studied a recording of your voice, you do not know the real sound of your voice. You are likely to be surprised!

▶ According to speaking consultant and author Bert Decker (1992) , ". . . the voice on the tape is much closer to what others actually hear than the voice we ourselves hear as we speak." The voice we hear is conducted largely through the bones in our head, while the voice others hear is transmitted through the air. Decker notes that the "reel" voice is the "real" voice, at least as far as audiences are concerned.

> *We are all public speakers. There's no such thing as a private speaker— except a person who talks to himself.*
> Bert Decker

▶ Unless you plan to talk only to yourself, you ought to know how you sound to others. If you don't like it, change it.

▶ Record all or part of your next presentation as part of your preparation to speak. A small microcassette recording device unobtrusively placed on the lectern or otherwise near you will capture the verbal and vocal components of your presentation. Privately study the recording, identify strengths and weaknesses, and explicitly build on the former while you diminish the latter.

Memorize a few, selected portions of your presentation

▶ The opening statement, to create a strong beginning.

▶ The closing statement, to ensure that the intended message has been communicated and to transition to questions and discussion.

▶ Quotations, to ensure you repeat them accurately.

Visualize your complete presentation and the audience's response

▶ Assemble a profile of the likely audience by conferring with organizers of the event, visiting the sponsoring organization's website, and using your network. Learn enough about the audience so that you can approach them from their perspective. Include items such as the following in the audience profile:
- Likely number
- Disciplines/specialties represented
- Education
- Age
- Years of practice/work
- Gender
- Reasons to be present

▶ Based on the audience profile, identify one or more ways to connect with all or portions of the audience near the beginning of your presentation. Perhaps you share a professional discipline, have worked on or are working on similar projects, or have lived in their geographic area.

▶ Learn what you can about the room or space within which the presentation will occur. If feasible, visit the room well before the actual presentation to learn firsthand about it and possibly practice your presentation out loud. Items of interest include audience seating arrangements (e.g., classroom, U-shaped), presence of a stage or platform, location of the screen relative to the audience, and availability of a remote control if you are using PowerPoint or a similar graphics system.

▶ With information like the preceding in hand, practice your presentation out loud again, this time visualizing where you will be relative to the audience and what you will be doing to enliven communication.

▶ Practice moving about the room, making gestures, changing voice cadence and volume, pausing, using props, and speaking directly to particular members of the audience, other speakers, and the moderator. Wear the type of clothing you are likely to wear when you actually speak.

Practice does not make perfect. Only perfect practice makes perfect.
Vince Lombardi

▶ In a sense, the entire room is your stage—at least while you are speaking. You are the producer, director, and actor. Use visualization to fully utilize all verbal, vocal, and visual communication channels.

Ask a spouse, family member, friend, or colleague to attend your actual presentation and provide you with a frank critique

- ▶ Encourage them to watch and listen to you while observing the audience. Ask them to listen to what audience members say among themselves during and after your presentation. Request that they record audience questions.

- ▶ What were your strengths?

- ▶ What needs improvement?

- ▶ Commit to building on your strengths and making improvements.

Read the following related lesson

- ▶ Lesson 22, "P⁵: Preparing, Presenting, and Publishing Professional Papers"

Study the source cited in this lesson

- ▶ Decker, B., and J. Denney. 1992. *You've got to be believed to be heard.* New York: St. Martin's Press.

Refer to the following supplemental source

- ▶ Urban, H. 2003. *Life's greatest lessons: 20 things that matter.* New York: Simon & Schuster.
  Chapter 11, "Real Motivation Comes from Within," urges us to create mental pictures of the success we desire, noting that we don't think in words; we think in pictures.

Subscribe to one or more of these e-newsletters

- ▶ *Abbott's Communication Letter* is a free monthly e-newsletter from Abbott Management Services that provides speaking tips. Recent topics include the "everybody knows" syndrome and the opportunity cost of communication. To subscribe, go to http://www.abbottletter.com.

- ▶ "Great Speaking" is a free e-newsletter produced twice per month by Antion & Associates. Offers very detailed speaking tips. To subscribe, go to http://www.antion.com.

- ▶ "The Professional Speaker" is a free e-newsletter from Bill Brooks that offers specific speaking tips. Example subject areas are use of humor, caring for the audience, and telling stories. To subscribe, send an e-mail to bill@thebrooksgroup.com.

Visit one or more of these websites

- ▶ "Advanced Public Speaking Institute" (http://www.public-speaking.org) is the website of the Advanced Public Speaking Institute. Available free speaking resources include checklists (e.g., a detailed pre-program questionnaire), a large public speaking glossary, and over 100 articles and links to other websites.

▶ "National Speakers Association" (http://www.nsaspeaker.org/) is maintained by the NSA. This organization "provides resources and education to advance the skills, integrity, and value of its members and the speaking profession." Included on the website are information on joining the organization, coming events, and a free knowledge bank and resource center that you can search using keywords or phrases.

▶ "Presentation Skills, Public Speaking and Professional Speaking" (http://www.antion.com) is the website of Antion & Associates. Very commercial with numerous materials and items for sale. Also offers free articles.

▶ "Toastmasters International" (http://www.toastmasters.org/) is the website of Toastmaster's International, which has the tagline "making effective communication a worldwide reality." Included are free speaking tips, access to an online store offering a wide variety of products, and information on how to find a Toastmaster's International Club near you.

The notes I handle no better than many pianists.
But the pauses between the notes—ah, that's where the art resides.
Arthur Schnable

24

DAD Is Out, POP Is In

> *With public sentiment, nothing can fail; without it nothing can succeed.*
> *Consequently, he who molds sentiment goes deeper than*
> *he who enacts statutes or pronounces decisions.*
>
> *Abraham Lincoln*

Two major challenges face today's manager of an infrastructure/environmental project within the public arena. The first is finding solutions to increasingly complicated technical problems. The second is communicating effectively with the public, recognizing both the public's increasingly elevated goals and the public's growing understanding of science and technology. Many and varied stakeholders want to be involved in infrastructure/environmental project decisions and should be given the opportunity to do so.

Unfortunately, some public works personnel, engineers, and other technical professionals fail to appreciate the importance of the communication challenge, or they recognize the challenge but are not prepared to meet it. The traditional **decide-announce-defend** (DAD) approach is no longer appropriate. The progressive and inclusive **public owns project** (POP) view is more likely to be effective given the changing nature of the public's expectations and knowledge.

An infrastructure/environmental effort that fails to include a public interaction program plans to fail. At the outset of projects, engineers and other technical professionals are clearly in a position to initiate, or strongly suggest the initiation of, a public interaction effort. But we often don't, or we do it half-heartedly, or start too late. Why? Professor Edward Wenk (1996) theorizes that "...by inclination and preparation, engineers approach the real world as if it were uninhabited. Yet...everything they do is for people." Janet C. Herrin and Arland W. Whitlock, program managers with the Tennessee Valley Authority, suggest, somewhat harshly but perhaps accurately, that the cause of some communication failures lies with the formal and informal education engineers, and perhaps other professionals, receive. According to them,

> Engineers are taught very few skills in interpersonal relationships, much less those of public interface and involvement. We spend little, if any, time addressing it at our conferences and conventions. We then spend thousands of hours and millions of dollars defending our projects when threatened by delays and possible blockage by public intervention.

Many subgroups with very different, often competing agendas typically constitute "the public" or stakeholders. Examples of subgroups are environmental organizations, recreation clubs, service groups, professional societies, business associations, educators, students of all ages, individual citizens, and appointed and elected government officials at all levels. The success of a public interaction program is determined more by the number of different "publics" that participate than by the total number of individuals involved. Breadth of public representation and involvement is crucial.

Engineers should be especially wary of the temptation to exclude what we regard as "extremist" elements from the public interaction effort. These groups have a right to be part of the process and to express their views. Attempts to exclude them are likely to aggravate matters and precipitate or elevate conflict. In addition to affording them their rights, inclusion of extremists may lead to moderation of their positions as they are gradually exposed to new information and as they interact with other, more moderate spokespersons for other segments of the public.

Public opinion is the best judge of who's right and who's wrong.
Chinese proverb

The importance of conducting the public interaction effort throughout the program must be emphasized. Engineers should repeatedly interact with various stakeholders beginning on day 1 and extending throughout the process. The essence of effective public interaction effort is a carefully designed set of programs and events.

A public works effort that fails to plan for stakeholder interaction plans to fail. In other words, DAD is out, POP is in (Walesh, 1999). Savvy managers and leaders proactively work with stakeholders.

Suggestions for Applying Ideas

Design a public interaction program that will achieve these three objectives (Walesh, 1993, 1999)

> ▶ *Demonstrate empathy and concern*. Demonstrate to the public and their appointed and elected representatives that we professionals are aware of the problems, at least in a general sense; we want to learn more about them; and

we want to seek solutions. The public's position in the early part of a planning process might be represented by the statement, "I don't care how much you know until I know how much you care."

▶ *Gather supplemental data and information* pertinent to the public works effort. Interested citizens and officials, if informed about what they believe to be a potentially useful effort, are likely to contribute ideas on solutions and useful data and information. For example, I've participated in many stormwater and floodplain management projects. Citizens affected by these projects, once they were convinced that the project team was sincerely interested in helping them, typically provided a wealth of previously unavailable data and information, such as high water marks, photographs of flood damage, rainfall measurements, and possible solutions.

▶ *Build a base of support* for rapid project implementation. Enlightened citizens and officials—who have been informed about an infrastructure/environmental project and have been given an opportunity to participate in preparation of a plan—are likely to become supporters of the plan, to help interpret it for others, and to help implement it

Create an effective stakeholder interaction process by selecting appropriate programs, events, and other tools (Walesh, 1993, 1999)

▶ *Advisory committees.* Members who should represent various stakeholders might include the following: elected and appointed government officials; business and professional people from the private sector; regional, state, and federal officials; K–12 schoolteachers and university professors; and representatives of environmental and recreation groups.

▶ *Public meetings.* One or more introductory meetings should be held early in the planning process, primarily to help achieve the first of the three previously mentioned public interaction objectives; that is, earning the public's trust. One or more intermediate public meetings can focus on status reports and presentation of alternatives that are under consideration. Draft recommendations can be presented at a final public meeting. To the extent feasible, these meetings should be conducted within affected areas to facilitate ease of access and as a symbol of concern.

▶ *Contacts* with engineering and planning firms, land developers, and professional societies. Because of their expertise and influence, such entities and groups can provide valuable insight, useful data and information, and support during plan implementation.

▶ *Presentations* to service clubs and other community groups. Knowledgeable and influential community leaders are typically members of one or more civic organizations, such as service clubs, environmental groups, homeowner's associations, and professional associations.

▶ *Field reconnaissance* and contacts. One or more members of the professional team usually performs a field reconnaissance of the study area by vehicle and by walking, during which they may make notes, obtain measurements, and take photographs. This reconnaissance provides opportunities for informal, one-on-one interaction with inquisitive members of the public.

▶ *School programs.* By educating schoolchildren about environmental, economic, and other issues, a two-fold result can be achieved. The students gain understanding and, to the extent they share what they learned with their parents, the knowledge is disseminated.

▶ *Guided and self-guided tours.* Interested individuals and groups can be provided with guided tours of large-scale projects. A single bus or van, preferably equipped with a public address system, should be used for a guided tour so that all participants can easily travel together and be provided with an informative narrative between stops. Self-guided tours are also possible if a written tour guide is available.

▶ *Briefings* for newly appointed or elected public officials. By being introduced to issues and being provided with basic information on ongoing or completed planning or design efforts, new public officials are more likely to be supportive, especially when funding decisions need to be made.

▶ *Workshops.* Professionals can conduct workshops for interested citizens and public officials. These events provide an opportunity for in-depth exploration of substantive topics such as issues, findings, alternatives, recommendations, and operations.

▶ *Electronic-based access and input.* Many options are available, such as toll-free telephone numbers, recorded messages, websites, and e-mail.

▶ *Clean-up projects.* Government agencies or units that are responsible for environmental projects can draw attention to and obtain support for the effort by encouraging and supporting the organization of clean-up efforts by environmental groups or individual citizens. Educational information, as well as materials and equipment—such as protective clothing, tools, bags and other containers, and trucks—can be provided by the government units or agencies.

▶ *Negotiated conflict resolution* among contending interests. Contending interests—which might include environmentalist citizens, land developers, and government entities—can be provided with an opportunity to negotiate an acceptable solution provided it satisfies applicable rules and regulations.

Read the following related lessons

▶ Lesson 15, "So, What Do You Know about Bluebirds?"

▶ Lesson 16, "Talk to Strangers"

▶ Lesson 34, "TEAM: Together Everyone Achieves More"

▶ Lesson 49, "Giving to Our Profession and Our Community"

Study one or more of the following sources cited in this lesson

▶ Herrin, J.C., and A.W. Whitlock. 1992. "Interfacing with the public on water-related issues: what TVA is doing." *Saving a threatened resource*. New York: ASCE, pp. 293-298.

▶ Walesh, S.G. 1993. "Interaction with the public and government officials in urban water planning." *Hydropolis—the role of water in urban planning*. Proceedings of the International UNESCO-IHP Workshop. Leiden, The Netherlands: Backhuys Publishers.

▶ Walesh, S.G. 1999. "DAD is out, POP is in." *Journal of the American Water Resources Association* (June): pp. 535-544.

▶ Wenk, E., Jr. 1996. "Teaching engineering as a social science." *The bent of Tau Beta Pi* (Summer), pp. 13-17.

Refer to one or more of the following supplemental sources

▶ Eschenbach, R.C., and T.G. Eschenbach. 1996. "Understanding why stakeholders matter." *Journal of Management in Engineering* (Nov./Dec.), pp. 59-64.

Advises engineers that what may seem logical to them may not be acceptable in today's regulatory and political circumstances. Offers this very demanding advice: "For a project to be carried out in the current political climate, almost all interested stakeholders must feel that the benefits of a proposed project outweigh the perceived risks." Presents case studies and draws lessons learned from them and other sources.

▶ Farris, G. 1997. "Citizens advisory groups: the pluses, the pitfalls and better options." *Water/Engineering and Management*, (Oct.), pp. 28-31.

Includes suggestions on when and how to form citizens' advisory groups and how to facilitate their work.

Engineers must be society-wise as well as technology-wise.
Warren J. Viessman, Jr.

Learning and Teaching

Ideally, each of us would, on a daily basis, learn and teach. Most of us have the ability to do both. All of us have the responsibility to do the latter, in the spirit of helping others by sharing what we know, and the need to do the former. Success in a science- and technology-based profession like engineering depends on each individual's and each organization's ability to keep up with new developments. In addition, we share with all professions the need to study and understand changing social, economic, environmental, and political conditions.

This section offers suggestions for individual learning, one-on-one learning and teaching, and group efforts. The underlying theme: our responsibilities as managers and leaders include learning and teaching.

25

Professional Students

Learning is a treasure that will follow its owner everywhere.
Chinese proverb

While I was still in graduate school and many of my contemporaries were accepting "real jobs," I was increasingly labeled a "professional student" (among other things, such as egghead, dweeb, bookworm, nerd, geek, freak, and wonk). I didn't mind very much, as I had always liked studying and learning; and I applied positive monikers to myself, like studious, serious, good student, and thinker. Near the end of my studies, my grandmother Emma Wolcott, who never had an unpleasant word for me and had left school after the fifth grade, said, "Stuart, what a shame! You are 27 years old and not working." I was so embarrassed! But I persisted, and I am now more than twice that age and still working—if you call writing and independent consulting a real job! The need to continue learning has never ended. I am still a professional student, but instead of feeling embarrassed, I feel fortunate!

Two aspects of our technical professions give added impetus to the need for lifelong learning. The first is that our science and technology foundation changes rapidly. Ongoing study is needed to stay current. Second, the non-technical side of engineering and other technical professions evolves. Consider, for example, the rise or prevalence of the following in the past decade alone: total quality management (TQM), privatization, reengineering, empowerment, acquisitions, corporate universities, outsourcing, knowledge management, virtual teams, the web, teleworking, distance learning, e-commerce, and environmental regulations.

We can continue to be "professional students" by implementing the following three learning mechanisms:

1. Meet new work challenges with as-needed, on-the-go learning. This just-in-time, just-enough learning involves conferring with a co-worker,

searching your employer's knowledge management system, tapping into your personal network, reading a book, or going to the World Wide Web.

2. Enroll in classes and attend seminars.

3. Become actively involved in technical and other professional societies, for example, by serving on a technical committee or presenting a paper.

For each of the three learning mechanisms, we are in charge. Employers may support or even partner with us, but ongoing learning is ultimately a self-directed enterprise. Prolific author Isaac Asimov agrees, saying, "Self-education is, I firmly believe, the only kind of education there is. The only function of a school is to make self-education easier; failing that, it does nothing."

> *How many a man has dated a new era in his life from reading of a book.*
> Henry David Thoreau

Being a fully functioning engineer means many things, one of which—and near the top of my list—is being a professional student. Our profession demands continuous learning. This is not a problem but an opportunity—an opportunity to exercise and renew our minds. What are you studying?

Suggestions for Applying Ideas

Take a strong lead in creating a balanced personal education and training program for yourself; nobody else will!

▶ Very few engineering employers maintain systematic, ongoing education and training programs (*Engineering Times*, 1992).

▶ For example, a survey of 165 A/E firms indicated that "training is typically conducted on a hit-or-miss basis, is rarely documented, and is frequently evaluated."

▶ The survey also indicated that only 8% of the firms operate formal training programs and only 14% have a manager of in-house training.

> *I will study and get ready and someday my chance will come.*
> Abraham Lincoln

▶ An observation based on my experience as an education and training consultant: Many, if not most, engineering businesses view education and training as a cost, not an investment. Accordingly, if business falters, the already modest education and training efforts are usually temporarily reduced or eliminated. The irony is that knowledge, skill, and attitude deficiencies may be the cause of the organization's downturn and could be remedied by focused education and training.

Design your personal education and training program to be aligned with your chosen roles and to help you achieve your related professional and other goals

▶ If you've articulated SMART goals and have documented them in writing, as suggested in Lesson 3, you should be able to advance many of your goals by obtaining related knowledge and skills.

In every person that comes near you look for what is good and strong; honor that; try to imitate it, and your faults will drop off like dead leaves when their time comes.

John Ruskin

▶ Assume, for example, that one of your multi-year goals is to profitably and otherwise successfully manage a design project with a contract amount of $200,000 or more. Then a possible education and training step is to study for the Project Management Professional (PMP) examination administered by the Project Management Institute (http://www.pmi.org/).

▶ View your work environment as a classroom with many and varied learning resources (Hensey, 1999):

- Colleagues, supervisors, and other personnel at your level and above and below you in the organizational hierarchy.
- Clients, customers, and regulators.
- Suppliers, vendors, and business partners.
- Outside consultants and other experts.
- Books, journals, magazines, videotapes, CDs, internal and external seminars, workshops, and conferences.

Study the lives of accomplished engineers to prepare yourself for achievement

▶ *Octave Chanute* (1832–1910), a versatile civil engineer who concentrated on railroads for most of his career. However, in the latter part of his career he studied and experimented with gliders, the results of which influenced the Wright brothers (Weingardt, 2001).

▶ *Fred N. Severud* (1899–1990), an innovative civil engineer whose achievements included design of the Gateway Arch in St. Louis. He also authored several books and was active in religious work (Weingardt, 2002).

▶ *Charles P. Steinmetz* (1865–1923), an electrical engineer who came to the U.S. in 1889 from Germany and led development of alternating current theories that enabled expansion of the power industry in the U.S. To support electrical engineering education, he published several textbooks and, later in his career, helped found and then headed the School of Electrical Engineering at Union College. Afflicted at birth with the physical deformity of hunchback, Charles Steinmetz was not to be denied; he had a remarkable career (http://chem.ch.huji.ac.il, September 14, 2003.

Not all readers become leaders. But all leaders must be readers.

Harry S Truman

▶ *Theodore von Karman* (1881–1963), born in Hungary, studied mechanical engineering and related topics and earned a doctorate in Europe, and soon came to the U.S. His project work included wind tunnels, Zeppelins, rockets, gliders, jet airplanes, and helicopters. He founded and led laboratories in Europe and the U.S. and he made many contributions to fluid mechanics theory (http://aerodyn.org, September 14, 2003).

Use your resume as a tool for monitoring your progress as a "professional student"

▶ Assuming you haven't updated your resume for several months, use it to answer these questions:

- What new concepts, ideas, or principles have you studied and applied?
- What new areas of technology have you mastered?
- What new management and/or leadership techniques have you learned or used?
- What new skills have you acquired and used?
- What new knowledge have you shared with professional colleagues via presentations and publications?
- What are you doing today that you hadn't known about or thought about doing months or a year or so ago?

▶ The answers to the preceding questions should prompt you to update your resume. However, if your resume doesn't require updating, you probably do.

Read the following related lessons

▶ Lessons 25 through 30 in Part 3, "Teaching and Learning"

▶ Lesson 3, "Smart Goals"

▶ Lesson 45, "Our Most Important Asset"

▶ Lesson 52, "Looking Ahead: Can You Spare a Paradigm?"

Study one or more of the following sources cited in this lesson

▶ *Engineering Times*. 1992. "A/E firms hit or miss on training" (Feb.).

▶ Hensey, M. 1999. *Personal success strategies: developing your potential*. Reston, Va.: ASCE Press.

▶ Katz, Eugenii, "Charles Proteus Steinmetz," http://chem.ch.huji.ac.il/~eugeniik/history/steinmetz.html (accessed September 14, 2003). (Information about Charles P. Steinmetz.)

▶ Filippone, Antonio, "Theodore von Kármán (1881–1963)" http://aerodyn.org/People/vonKarman.html (accessed September 14, 2003). (Information about Theodore von Kármán.)

▶ Weingardt, R.R. 2001. "Engineering legends: Octave Chanute and George Ferris." *Leadership and Management in Engineering* (Oct.), pp. 88-90. (This regular column provides valuable information about and ideas on engineering leaders.)

▶ Weingardt, R.R. 2002. "Engineering legends: Roland C. Rautenstraus and Fred N. Severud." *Leadership and Management in Engineering* (Jan.), pp. 44-56.

Refer to the following supplemental sources

▶ Billington, D.P. 1996. *The innovators: the engineering pioneers who made America modern*. New York: John Wiley & Sons.

As stated by the author in the Preface, this "book treats U.S. engineering history as an interplay of three perspectives: what great engineers did, the political and economic conditions within which they worked, and the influence that these designers and their works had on the nation." He goes on to say, perhaps surprisingly, "We shall discover that the essence of engineering lies not just in natural science, as is usually thought, but also in social science and the humanities."

▶ Fredrich, A.J. (ed.). 1989. *Sons of Martha: civil engineering readings in modern literature*. New York: ASCE.

Included in this collection of civil engineering readings are biographical sketches of civil engineers. John and Washington Roebling, father and son respectively, led, with the help of Washington's wife, Emily, the planning, design, and construction of the Brooklyn Bridge. Sanitary engineering educator Gordon M. Fair is described as "world-renowned for the ideas he shared rather than the projects he designed or constructed." The differences in the focus and admirable accomplishments of Roeblings and Fair are indicative of the wide variety of accomplished civil engineers described in this book and, much more profoundly, civil engineers found within the profession.

Visit this website

▶ "ClassesUSA" (http://www.classesusa.com/) helps find an online class or degree program to meet your needs.

It's only a mistake if you don't learn from it.

Richard G. Weingardt

Garage Sale Wisdom

The man who does not need good books has no advantage
over the man who can't read them.
Mark Twain

My wife enjoys browsing garage sales, and although I am much less enthusiastic, I often go with her. While she typically hunts for treasures, I head for the inevitable piles or boxes of old books. Our roles reverse in used bookstores: I enjoy cruising the stacks and my wife is a good sport by coming along.

Books available at garage sales and in used bookstores cover the gamut from those labeled "two years or younger" to encyclopedia sets that end with World War II, from romance novels to medical reference books. My targets tend to be biographies, histories, and management and leadership books. As a result of visits to garage sales and used bookstores, I have found some interesting books, all of which were, although this was a secondary issue, terrific bargains.

An Indiana garage sale yielded motivational author Napoleon Hill's *Think and Grow Rich*, published in 1960. I had heard of but never read this enlightening book. Contrary to the narrow monetary theme suggested by the title, Hill explains how he searched for the secret that enabled a few individuals to achieve great significance and success, financial and otherwise. The secret, based on Hill's 25-year study, is the power of visualization and the subconscious mind.

Management and Leadership, authored by Carl F. Braun, president of a manufacturing firm, was purchased for a nickel at a Florida garage sale. Consider some thoughts from this book: "A company cannot rise above its people." Teamwork occurs when the leader practices "teaching, helping, guiding, encouraging" in contrast with using "decree, dictate, command." "Ethics is not a trimming to the business tree. It is a mighty root of it." "Few leaders fully grasp the meaning of the verb, to lead, . . . to initiate, . . . to instruct and guide, . . . to take responsibility, . . . to be out in front." This book, which offers advice consistent with today's most enlightened thinking, was published in 1954!

Harper's Anthology for College Courses in Composition and Literature—Prose, was purchased in the used book section of a Florida shop. Included in this 1926 book are the maxims of the French moralist Francois La Rochefoucald. One example of his 17th century thoughts applicable to today is, "As it is the characteristic of great wits to convey a great deal in a few words, so, on the contrary, small wits have the gift of speaking much and saying nothing." And consider the appropriateness of this thought for engineers and the crucial nature of our judgment: "To know things well, we should know them in their details; but as their details are almost infinite, our knowledge is always superficial and imperfect." And this maxim for consultants: "We may give advice but we cannot inspire the conduct."

Other volumes in my library that were obtained from garage sales and old bookstores include the following:

- Dale Carnegie's *Public Speaking and Influencing Men in Business*, first published in 1926 and purchased at an Indiana garage sale.
- *The Human Side of Enterprise*, published in 1960 by Douglas McGregor and found in a New Jersey used bookstore.
- *Essays* by Ralph Waldo Emerson, published in 1936 and found in a Florida shop.

Buying old books has benefited me in two ways. First, I have acquired useful skills and knowledge by reading the books, and my perspective has been expanded. For example, my discovery and subsequent reading of Napoleon Hill's *Think and Grow Rich* fleshed out some of my earlier intuitive thinking. Second, given that the books mentioned in this lesson were all originally published four or more decades ago, and that they offer observations and advice similar to that offered today, I am gradually concluding that the things that are really important change little over decades, if not centuries.

> *I keep to old books, for they teach me something; from the new I learn very little.*
>
> Voltaire

They say that what goes around comes around. Garage sale and similar older books in my library suggest a possible variation on the preceding: the tried and true, while they may be ignored, never go away. Reading old books helps distinguish the truly new from the superficial, the wheat from the chaff.

Suggestions for Applying Ideas

Purchase some "garage sale wisdom," skim or read it, and consider the following questions

▶ What's different between then and now, both with regard to the book's subject matter and about the social, economic, and political environment?

▶ Are the differences matters of substance or of minor significance?

▶ What new ideas and information did you find in the old book?

Take a new look at some of your old college texts and other books

▶ This suggestion assumes that you, like me, have difficulty getting rid of old books. Perhaps your old books have a place of honor in your office or den, or maybe they are stored and almost forgotten in the attic, basement, or garage. What do these "old friends" have to say today in contrast with what you recall from back then? Some possibilities:

- You now have an answer for some of those question marks you placed in the margins on various pages.
- A theory, introduced and stressed in one of your engineering courses, seemed irrelevant then but has repeatedly proved, in application, to be of great practical value in your work.
- An idea you read about back then in one of your liberal arts courses proved to be a turning point for you; it profoundly influenced the direction of your life.
- You realize that, although you thought you knew much back then, you didn't. You still don't know much relative to what you now understand about what there is to know.

Book: A garden carried in a pocket.
Arabian proverb

Read the following related lesson

▶ Lesson 27, "Read and You Won't Need a Management Consultant"

Study the following sources cited in this lesson

▶ Braun, C.F. 1954. *Management and leadership.* Alhambra, Calif.: C.F. Braun & Company.

▶ Carnegie, D. 1935. *Public speaking and influencing men in business.* New York: Association Press.

▶ Emerson, R.W. 1936. *Essays.* Reading, Pa.: The Spencer Press.

▶ Hill, N. 1960. *Think and grow rich.* New York: Fawcett Crest.

▶ Manchester, F.A., and W.F. Giese (eds.). 1926. *Harper's anthology for college courses in composition and literature.* New York: Harper & Brothers.

▶ McGregor, D. 1960. *The human side of enterprise.* New York: McGraw-Hill.

To add a library to a house is to give that house a soul.
Marcus Tullius Cicero

Read and You Won't Need a Management Consultant

A man is known by the company his mind keeps.
Thomas Bailey Aldrich

President Harry S Truman once said, "There is nothing new in the world except the history you do not know." In keeping with the spirit of his statement, I doubt that there is an employee, client, or stakeholder problem in any of our public and private technical organizations that hasn't occurred and been solved before.

My management consulting practice focuses on solving problems in engineering organizations that are caused by non-technical or "soft-side" deficiencies and not by technical shortcomings. After reflecting on my experiences, I have concluded that very few engineers and other technical managers read about management, leadership, and related topics. This strikes me as unfortunate for them, because the literature is rich with ideas and information on how each of us can assess our skills, articulate our aspirations, and create and implement a personal improvement action plan. Instead, many firms must, fortunately for me, retain the services of a management consultant to help them identify and resolve problems they easily could work on themselves.

My wife says that too much of my reading is work-related. Frankly, she's right: the books and other items I read are substantially management and leadership material, for two reasons. First, I enjoy the stimulation of trying to sort out old ideas, old ideas gussied up to look like new ideas, and occasional truly new ideas. Second, effectiveness in my consulting work requires that I stay current.

Are you inclined to read more widely in and beyond the management and leadership area? If so, I respectfully suggest that you consider some of the following:

- *Think and Grow Rich* by Napoleon Hill (1960). This old book has a somewhat misleading title. That is, the book does not focus on accumulation of material wealth but does stress achieving ambitious aspirations. Hill's research into the lives of highly accomplished people, as detailed in his book, convinced him of the power of visualization in combination with the workings of the subconscious mind.

- *As A Man Thinketh* by James Allen. This thoughtful little book argues, consistent with Napoleon Hill's thesis, that we become what we think, positive or negative, through the power of our subconscious.

- *The 7 Habits of Highly Effective People* by Stephen Covey (1990). Unlike many of the gimmicky self-help books, Covey focuses on application of sound personal and interpersonal principles to achieve win-win outcomes.

- *Leadership and Management in Engineering*, a quarterly publication of the American Society of Civil Engineers, and its continuing predecessor, *Journal of Management in Engineering*, which is published every other month. Issues typically are loaded with management and leadership topics presented in a variety of formats, ranging from short tidbits to in-depth, peer-reviewed articles.

- *First You Have to Row a Little Boat* by Richard Bode (1993). Sailing is used as an insightful analogy to life.

- *Discovering the Future: The Business of Paradigms* by Joel Arthur Barker (1993). Described here are the dangers of paradigm paralysis and the possibilities inherent in a paradigm pliancy perspective.

- *Engineering Your Future: The Non-Technical Side of Professional Practice in Engineering and Other Technical Fields*, 2nd Ed. , by Stuart G. Walesh. Excuse the immodesty of offering my own book. However, this combination text and reference book targets and is totally devoted to helping the young technical professional develop the non-technical or "soft-side" skills necessary to be an effective manager and leader.

> *Never attribute to malice what can be explained by incompetence because incompetence really is so much more common than deliberate malice.*
> Josef Martin

The answers to your personal, project, and organizational managing and leading challenges are "out there." One of the best ways to obtain those answers is to read deeply and broadly.

Suggestions for Applying Ideas

Follow Mortimer Adler's advice on "how to read a book" (Grugal, 2002a)

▶ Take an active role in your reading. For example, as you read, frequently answer this question in your own words: "What is really being said?" Then consider questions like "Is it true?" and, if so, "How can I use it?"

▶ Take notes, perhaps in the book's margins. Include ideas, information, questions, and action items.

> He that reads and grows no wiser seldom suspects his own deficiency, but complains of hard words and obscure sentences, and asks why books are written which cannot be understood.
>
> *Samuel Johnson*

▶ Test yourself. As a result of reading the book, what new ideas and information did you acquire and, more importantly, how will your thinking and behavior change as a result?

Read widely and eclectically

▶ Most of the readings suggested in this lesson are not explicitly connected to managing and leading.

▶ This implies the power of eclectic reading, that is, occasionally reading books, periodicals, newspapers, and other materials drawn from outside of our normal reading patterns and not necessarily related to our work. Consider reading biography, history, and philosophy. Eclectic reading also means reading articles and book with points of view different from yours.

▶ We can also eclectically view television, listen to the radio, and attend the theater. However, the advantage of eclectic reading is that we can stop and ponder what we read; we can control the pace.

> All things are filled full of signs, and it is a wise man who can learn about one thing from another.
>
> *Plotinus*

▶ If you don't normally do so, occasionally read newspapers such as the *Christian Science Monitor, Financial Times, Investor's Business Daily, Wall Street Journal,* and magazines and other periodicals such as *Fortune, Money, National Review,* and *Economist.*

▶ Eclectic reading may provide a fresh perspective on tried and true processes, offer a glimpse of future technologies and service needs, expand your vocabulary, provide added insight into human nature, and introduce you to potential business partners.

Read the following related lessons

▶ Lesson 25, "Professional Students"

▶ Lesson 26, "Garage Sale Wisdom"

Study one or more of the following sources cited in this lesson

▶ Allen, J. 1983. *As a man thinketh.* Marina Del Ray, Calif.: DeVorss & Company.

▶ Barker, J.A. 1993. *Discovering the future: the business of paradigms.* St. Paul, Minn.: ILI Press.

▶ Bode, R. 1993. *First you have to row a little boat.* New York: Warner Books.

▶ Covey, S.R. 1990. *The 7 habits of highly effective people: restoring the character ethic.* New York: Simon & Schuster.

▶ Grugal, R. 2002a. "Reading is an art form." *Investor's Business Daily* (May 5). (Summarizes Mortimer Adler's 1940 book, *How to read a book.*)

► Hill, N. 1960. *Think and grow rich*. New York: Fawcett Crest.

► Martin, J. (pseudonym). 1988. *To rise above principle: the memoirs of an unreconstructed dean*. Urbana, Ill.: University of Illinois Press.

► Walesh, S.G. 2000. *Engineering your future: the non-technical side of professional practice in engineering and other fields*, 2nd Ed. Reston, Va.: ASCE Press.

Reading maketh a full man, conference a ready man,
and writing an exact man.

Francis Bacon

28

Caring Isn't Coddling

It's a funny thing about life;
if you refuse to accept anything but the best, you very often get it.
W. Somerset Maugham

*T*hinking back to my days as an engineering dean, I recall the gist of a telephone conversation I had with the mother of a high school student. She expressed skepticism about "church-related" educational institutions because she feared that their mission would lead to over-protection and under-stimulation of her child. The conversation remains vivid because my response was less than satisfactory, which troubled me. I should have said that while we did care, we did not coddle. On the positive side, the conversation stimulated my thinking, changed my approach, and led to this lesson.

As we develop our personal relationships, within and outside of professional work, many of us hope that a strong element of caring will be evident in our actions toward others and in their actions toward us. Caring, as used here, does not mean coddling. But if caring isn't coddling, what is it?

To answer this, think about those former teachers who really cared about you. They probably demonstrated their concern for you narrowly as a student and broadly as a person in the following kinds of ways: delivering well-prepared lectures, making regular and demanding assignments intended to deepen and broaden understanding of the course material, providing opportunities for independent study such as a research paper or laboratory project, encouraging you to participate in cocurricular and extracurricular leadership and service activities, offering an encouraging word at a discouraging time, and praising when nobody else seemed to notice what you had accomplished.

These meaningful interactions were not offered in a paternalistic, condescending, ostentatious manner; rather, these actions were part of a high expectations–high support environment intended to stretch without snap-

Give the other person a fine reputation to live up to.
Dale Carnegie

ping, provide example without expecting cloning, and build confidence without imparting arrogance.

Caring is also exemplified by the parent who says, "If it's worth doing, it's worth doing well," and by the colleague who, at a meeting, has the courage to ask the awkward question or raise the sensitive issue that almost everyone knows must be addressed. Caring is also shown by the manager who says no to the pleading employee who did not strive to meet the established requirements and now wants to avoid the adverse consequences.

You may also recall with disdain those teachers, supervisors, colleagues, and others who were generally "nice" but didn't expect all that much of you. Oftentimes you delivered in accordance with their expectations. You and they could have done so much more. Perhaps they didn't really care about you, or even themselves.

What is honored in a country will be cultivated there.
Plato

Caring isn't coddling. Caring is pushing, pulling, admonishing, stretching, demanding, encouraging, urging, challenging, and cajoling. Caring is high expectations coupled with high support. Caring helps individuals and organizations meet their goals and realize their full potential.

Suggestions for Applying Ideas

Recall someone who, during a formative period in your life, consistently had high expectations for you and matched those expectations with high support

▶ How did they communicate their expectations?

▶ How did they offer support?

▶ How did you benefit?

▶ What did you learn from that relationship that might enable you to, in turn, offer high expectations–high support to others?

Pass along the benefits of high expectations–high support

▶ Select a supervisee who would benefit from structured guidance.

▶ Initiate the kind of high expectations–high support experience that enriched your life.

▶ Articulate your and/or your organization's expectations for the selected person.

▶ Pledge your and the organization's support if the supervisee strives to meet the stated expectations or some mutually agreeable variations.

▶ Hold the selected person accountable to develop as expected, and hold yourself and your organization accountable to provide support as warranted.

► After a few months, reflect on the success of your efforts. Did your supervisee benefit from your efforts like you benefited from the efforts of your coach or mentor?

Read the following related lessons

► Lesson 2, "Roles—Then Goals"

► Lesson 3, "Smart Goals"

► Lesson 33, "Delegation: Why Put Off Until Tomorrow What Someone Else Can Do Today"

► Lesson 45, "Our Most Important Asset"

---

*Time and money spent in helping men do more for themselves
is far better than mere giving.*

Henry Ford

29

More Coaching, Less Osmosis

The only irreplaceable capital an organization possesses is the knowledge and ability of its people. The productivity of that capital depends on how effectively people share their competence with those who can use it.

Andrew Carnegie

Consulting assignments provide me with an opportunity to objectively see vertical slices of private and public technical organizations—from entry-level engineers and other technical professionals up through middle and senior managers, to the CEO or other top-most positions. The more experienced personnel typically hold the higher positions. That is why they are where they are; their knowledge and skills are broad and deep.

Sometimes experienced personnel complain about the ineptitude of the more junior personnel, especially in non-technical areas. Senior personnel consider junior personnel to be poor communicators who can't manage projects profitably, lack marketing interest and skill, get bogged down in technical matters, fail to meet client expectations, lack visibility in the community, and have little business or political sense. Frequently, young, technically proficient engineers are not even told about their shortcomings and deficiencies in these non-technical areas.

Unfortunately, senior staff often have little if any explicit involvement in sharing what they know with their junior counterparts. A common attitude is that entry-level personnel will realize—perhaps through osmosis—that they have "soft-side" skill deficiencies and will learn how—also by osmosis—to resolve them. Some eventually do learn, but often this learning comes negatively through mistakes that are costly and painful for them, their employers, and the organization's stakeholders.

Most junior members of organizations do and should hunger for knowledge about the non-technical aspects of professional practice or government service. But they do not automatically absorb this knowledge through osmosis.

"Coaching" is an effective way to leverage the experience of an organization's principals and other upper echelon personnel (Walesh, 1997). Coaching means occasional, one-on-one focused interactions between a caring senior person and a receptive junior person. Senior and junior usually refer to age but also could refer to differences in experience level between two individuals. For example, a younger person with more experience in a specific technical area could coach an interested older person.

> *It is one of the most beautiful compensations of this life that no man can seriously try to help another without helping himself.*
>
> Ralph Waldo Emerson

Coaching is distinguished from mentoring, which typically requires a major, personal, ongoing effort over an extended period of time, perhaps a year or more. Coaching is accomplished intermittently during the course of ongoing projects and activities. It is not a separate training activity.

How can experienced professionals coach others? How will the senior individuals have time to fit this additional activity into already busy schedules? The answer is to look, in the normal course of work, for specific coaching opportunities that will take the receptive, junior technical professional up the learning curve for "soft-side" skills and will provide him or her with some management and leadership insights. The coaching situations should be tailored to the level, needs, and receptivity of the young person.

Aspiring young engineers are usually quite bright, but they don't always know what they don't know, especially if it is outside of the technical or "hard-side" arena. They need to be told, by word and by example, that a judicious blend of technical and "soft-side" abilities is essential to personal and organizational success. Coaching is a highly leveraged way of sharing management and leadership skills and knowledge, and it can also provide a terrific return on investment for organizations in the private and public sectors.

Suggestions for Applying Ideas

Look for an opportunity to coach someone in your organization

▶ In your role as project manager, invite a young engineer to attend a client meeting and observe the proceedings. On the way to the meeting, describe what appear to be the key issues, how they were determined, and how they will be discussed with stakeholders. Outline the desired outcome of the meeting. After the meeting, jointly and constructively discuss the meeting and critique the tactics and skills of the various participants.

▶ In your role as city engineer, ask a junior engineer to attend a city council meeting at which the engineering department's annual budget will be presented and discussed. Prior to the meeting, explain the budget process, the basis for your department's budget, the role of elected officials and citizens, and the desired outcome. After the meeting, jointly analyze the ideas and information exchanged at the meeting and the decisions that were made.

▶ In your role as office manager or project manager, invest some quality time with a young professional by "walking" him or her through an in-process report on a project. Discuss and illustrate critical client-friendly features of the report, such as an attractive cover, a short-issues and bottom-line–oriented executive summary, documentation of all alternatives, ample white space, effective graphics, proper citation of sources, a list of abbreviations, and a glossary. Explain how the implementation of the project is likely to depend on a carefully crafted combination of the quality of the technical work and the manner in which it is presented in the report.

> *A coach is someone who can give correction without causing resentment.*
> John Wooden

▶ In your role as proposal and interview manager for a potential project, invite a young professional to sit in on an initial strategy session, some subsequent conversations with graphics and other personnel, a practice session prior to the actual interview, and the actual interview. Use these interactions as opportunities to share aspects of the corporate philosophy such as focusing on developing relationships with clients vis-à-vis pursuing projects. Emphasize the importance of client and project research and illustrate the advantages of teamwork as exemplified by clerical, engineering, graphic, and marketing personnel working together to identify and meet client and project requirements.

Recognize that some people who do things very well, like speaking to a citizen group or calling on a potential client or customer, don't know how they do it. Take action to learn from these individuals (Hensey, 1999).

▶ Accept that they cannot explain what they do and, therefore, that they will not be effective as active coaches. However, such high performers can serve as passive coaches.

▶ How? Arrange to observe those who do things well while they are doing it. Encourage the observers to identify the knowledge, skill, and attitudes that explain the admirable performance.

▶ Go one step further: "Interview" the exceptional performers as a means of learning more about how they do it. While active coaching may not be their forte, responding to questions may be.

Investigate mentoring, either as an informal effort or as an organizational program, if coaching proves successful in developing personnel

▶ Mentoring requires much more effort than coaching but, if done well, yields much greater benefits.

▶ Some definitions of mentoring:

- A relationship between a junior and senior staff member (in terms of age or experience) that primarily exists to support the personal and career development of the junior person (Fairchild and Freeman, 1993).
- Someone helping someone else learns something the learner would otherwise have learned less well, more slowly, or not at all (Bell, 1996).
- One person invests time, know-how, and effort in enhancing another person's growth, knowledge, and skills, to prepare the individual for greater productivity or achievement (Shea, 1994).

▶ Caution: An organizational mentoring program requires a major time commitment. Allow at least a year for a trusting and productive relationship to develop between a mentor and the mentee or protégé. Many hours of quality time, consisting of mentor-mentee conversations and between-conversation efforts, will be needed during that year.

▶ Bonar and Walesh (1995, 1998) provide a detailed case study of a successful mentoring program, and ASCE (no date) offers concepts and ideas

Read the following related lessons

▶ Lesson 45, "Our Most Important Asset"

▶ Lesson 47, "Eagles and Turkeys"

Study one or more of the following sources cited in this lesson

▶ ASCE. No date. *Mentors and mentees.* Reston, Va.: ASCE.

▶ Bell, C.R. 1996. *Managers as mentors.* San Francisco, Calif.: Koehler Publishers.

▶ Bonar, R.L., and S.G. Walesh. 1995. "Ownership transition: a mentoring case study." Proceedings of the Fall Conference of the American Consulting Engineers Council. Washington, D.C.: American Consulting Engineers Council.

▶ Bonar, R.L., and S.G. Walesh, 1998. "Mentoring: an investment in people." Proceedings of the Fall Conference of the American Consulting Engineers Council. Washington, D.C.: American Consulting Engineers Council.

▶ Fairchild, F.P., and H.E. Freeman. 1993. "Establishing a formal mentoring program in a consulting engineering firm." Presented at the ASCE Engineering Management Conference, Denver, Colo. (Feb.).

▶ Hensey, M. 1999. *Personal success strategies: developing your potential.* Reston, Va.: ASCE Press. Chapter 5, "Core Competencies for Living," and Chapter 6, "Learning How To Learn."

▶ Shea, G.F. 1994. *Mentoring: helping employees reach their full potential,* AMA Management Briefing, New York: American Management Association.

▶ Walesh, S.G. 1997b. "More coaching, less osmosis." Editor's Letter. *Journal of Management in Engineering* (July/Aug.).

Visit this website

▶ "International Mentoring Association," (http://www.wmich.edu/conferences/mentoring/genrinf.html). The IMD "exists to facilitate individual growth and development through best practices in mentoring." Included are articles from the Association's newsletters and links to other mentoring sites.

He has a right to criticize who has a heart to help.

Abraham Lincoln

Education and Training: From Ad Hoc to Bottom Line

In times of change, learners inherit the earth, while the learned are beautifully equipped to deal with a world that no longer exists.

Roland Barth

Clients and customers increasingly demand services that meet their needs. Citizens and the business community also expect more of all levels of government. Serving clients, customers, and the public competitively, especially in technically based areas, requires the service or product provider to have current knowledge and skills in both technical and non-technical areas.

A major portion of the knowledge assets of a consulting or manufacturing firm or government entity are what its employees know. This is intellectual capital. It is an asset that goes down the elevator or out the door at the end of the day. We are increasingly in the business of providing knowledge.

An investment in knowledge pays the best interest.

Benjamin Franklin

Maintaining and building intellectual assets are essential for the long-term profitability of a consulting or manufacturing firm and for the long-term viability of academic units and government entities. Government entities, for example, are increasingly challenged by various forms of privatization, a challenge that may lead to improvement if personnel receive reengineering education and training (E&T).

As a result, during the past decade or so, E&T programs of many organizations—including major manufacturers, some consulting engineering firms, and a few government entities—have undergone major changes. They have shifted from an ad hoc, individually focused model to a planned, organizationally focused model. In a few cases, the organizationally focused approach includes establishment of a corporate university or similar, highly structured operation or business unit. Some contrasts between the "old," traditional, ad hoc model of E&T and the "new," organizationally focused model follow (Roesner and Walesh, 1998):

- **Old**: Most learning is done in offsite and onsite seminar settings.
  New: Learning is accomplished internally and externally in a variety of ways, with the delivery mechanism selected on the basis of cost-effectiveness.

- **Old**: Employees in the main office have more E&T.
  New: Equal access to E&T is provided regardless of employees' geographic location.

- **Old**: Some participants obtain certificates.
  New: Organization is licensed to award continuing education units (CEUs).

- **Old**: Only employees participate.
  New: Stakeholders, such as clients, business partners, and vendors, are invited to participate as learners and teachers.

- **Old**: Organization pays all.
  New: Partnership between employee and organization.

- **Old**: Small fraction of personnel participate.
  New: Everyone is expected to participate.

- **Old**: Employee selects subject matter.
  New: Organization selects or approves subject matter based primarily on relevance to its vision, mission, and business or operating plan.

- **Old**: Some E&T is viewed as well-earned rest and relaxation.
  New: E&T is part of the business and expected to provide a return on the investment.

There are two likely negative reactions to the suggestion of creating more structured, organizationally focused E&T programs: First, that a business-oriented E&T program would "cost too much," and second, that employees would participate in the E&T and then leave, taking the investment with them. With respect to the first objection, experience indicates otherwise. Instead of spending more money, the intent is to leverage current expenditures more wisely. With respect to the second objection, experience also indicates otherwise as suggested by this anonymous thought:

> The only thing worse than educating and training people and having them leave, is not educating and training them and having them stay.

Suggestions for Applying Ideas

Determine which of these ten potential benefits of an organizationally focused E&T program could be applicable to your business or government entity (Roesner and Walesh, 1998)

1. Articulate and share the organization's culture, that is, its history, values, mission, and goals.

2. Attract and retain even higher quality personnel. A dominant characteristic of top-flight personnel is their desire for continuous learning.

3. Achieve more effective use of monetary and time resources currently being expended on E&T. This may be accomplished by focusing on the organization's needs (not individual "wants"), using many and varied delivery mechanisms tailored to the learning situation, reducing duplication of efforts, and holding individuals accountable for sharing and/or using what they know.

4. Close skill gaps, that is, shrink the differences between what personnel know and what they need to know, or know better, in order to achieve the organization's mission, vision, and goals.

5. Facilitate advancement of individuals, and the commensurate satisfaction and other rewards, by improving knowledge and skills.

> *Why, then, do companies manage it [human capital] so haphazardly? A principal reason, I believe, is that they have a hard time distinguishing between the cost of paying people and the value of investing in them.*
>
> *Thomas A. Stewart*

6. Grant CEUs, thus further recognizing employees; assisting them in meeting license, registration, and renewal criteria; and enhancing the stature of the organization.

7. Improve individual and corporate productivity and project fiscal performance, leading to improved profitability.

8. Improve marketing efforts by increasing understanding of the marketing process, enhancing marketing skills, and involving clients and potential clients in joint learning with the organization.

9. Enhance technical quality as indicated primarily by meeting client needs and other project requirements.

10. Hold E&T efforts to current costs, at least in the first year or so of operation. The thinking here is that one way to "sell" organizational managers and leaders on a new approach to E&T is to indicate that the monetary expenditures and time cost would not increase, but effectiveness would.

Find the optimum mix of sources and delivery mechanisms for providing E&T lessons, modules, and experiences

▶ While in-house workshops and seminars are likely to be one form of teaching and learning, other potential content sources and delivery mechanisms are:

- Carefully selected books, papers, and articles studied as part of one's personal professional plan, read as pre-module attendance assignments, or used as the basis for discussion groups.
- Audio and audio-visual cassettes used individually or in a group setting.
- Computer-assisted, nonlinear, interactive, individualized learning.
- Web-based interactive, individual, and group distance learning (synchronous and asynchronous).
- Correspondence courses.
- Conference calling, such as sessions at which project managers exchange "tips" or computer modelers discuss special applications.

- E-mail sharing, that is, a mechanism for announcing knowledge and information needs, on the assumption that someone will be able to help, and sharing knowledge and information, on the assumption that someone else may be interested.
- Attendance at external workshops, seminars, and conferences sponsored by professional or business organizations with the expectation to "share" knowledge (e.g., brown bag, memorandum) upon return to the office.
- Brown bag presentation of knowledge gained "on the job" or at a workshop, seminar, or conference.
- College/university courses taken on campus.
- College/university courses taken remotely.
- Live, multi-location, interactive audio/video instruction originating from within or outside of the organization's offices.
- Mentoring, that is, an extended one-on-one confidential relationship focused on meeting the mentee's needs.
- Tutoring or coaching, that is, special help from the organization's personnel or outside experts in mastering prescribed skills or material. "Outside expert" could mean consultant, joint venture partner, subcontractor, vendor, regulator, or client.
- Rotating personnel to encourage learning and cross-training.
- Preparing, using, and continuously improving written guidelines (sometimes called best practices, procedures, tips, or checklists) for frequently used technical and non-technical processes. Written guidelines are initially drafted by experienced personnel and then frequently updated and refined by other personnel who use the guidelines. While an initial major effort is needed to prepare written guidelines, long-term benefits include E&T, increased productivity, continuous improvement, elimination of valueless activities, and reduced errors and omissions.

Native ability without education is like a tree without fruit.
Samuel Johnson

Given the precarious nature of E&T programs in many organizations, a champion should defend and support the E&T program. This may be in addition to someone, perhaps from human resources, to administer the E&T program on a part-time or full-time basis.

▶ From the perspective of public and private organizations, E&T is often viewed as an expense to be cut whenever fiscal, time utilization, or other difficulties arise.

▶ Assume the organization is truly committed to E&T. Then a highly placed champion can help to bridge difficult times by being a knowledgeable advocate for the E&T program within the highest levels of the organization.

▶ The credibility of the organization's principals who advocate E&T will be tested during difficult times. Personnel will judge the organization's commitment to E&T by what the principals do, not by what they previously said.

Include on-going evaluation and support measures in the E&T program

▶ Continuously assess the quality of the program with emphasis on the extent to which newly learned material is being applied, or at least tried. Accountability will provide a return on the investment in E&T. Without accountability, the results will be disappointing. Consider using evaluation and support approaches such as the following:

- Pre- and post-E&T event quizzes to determine if new knowledge and skills are being acquired as a result of the events, although not necessarily used.
- End-of-workshop or other E&T event evaluations to obtain participant views on topics such as pre-event information and coordination, meeting room and other physical arrangements, usefulness of presented material, value of handouts, and presenter/facilitator effectiveness.
- Anecdotes that illustrate the successful application of newly acquired knowledge or skills. Use newsletters, meetings, and other means to share the anecdotes. While accounts of interesting incidents may not prove anything, their personal and specific nature raises awareness and encourages fresh thinking.

Are you green and growing, or ripe and rotting?

Ray Kroc

- Post-workshop or other E&T event evaluation by supervisors, conducted about two weeks after the event, to determine if personnel are at least trying to apply recently obtained knowledge and skills.
- Post-workshop or other E&T event evaluation by participants, completed about two weeks after the event, to determine if what was learned is being applied.
- Anonymous 360-degree evaluations of supervisors and managers to determine their support of and contributions to the organization's E&T effort.
- Trends in organizational indices, such as personnel retention, accidents, absence rates, liability claims, litigation, revenue per employee, leads generated, proposal success rates, client retention, recruitment costs, new clients served, stakeholder satisfaction, awards received, and profitability.
- Recognition of individuals who successfully complete E&T events or programs. For example, provide certificates of completion, award CEUs, send congratulatory letters to the homes of personnel, publish articles in organizational newsletters, and place notes in personnel files.

Read the following related lessons

▶ Lessons 25 through 30 in Part 3, "Teaching and Learning"

▶ Lesson 45, "Our Most Important Asset"

Study the following source cited in this lesson

▶ Roesner, L.A., and S.G. Walesh. 1998. "Corporate university: consulting firm case study," *Journal of Management in Engineering.* (March/April), pp. 56-63.

Refer to one or more of the following supplemental sources

▶ Farr, J.V., and J.F. Sullivan, Jr. 1996. "Rethinking training in the 1990's." *Journal of Management in Engineering* (May/June), pp. 29-33.

Surveys education and training trends as potentially applicable to the architectural, engineering, and construction industries. Advocates a new, more relevant and cost-effective education and training model that uses evolving electronic technology.

▶ Meister, J.C. 1994. *Corporate quality universities*. New York: Irwin Professional Publishing.

Argues that American businesses spend too little on education and training, spend it inefficiently, and focus too much on those who are already educated and trained. Advocates corporate universities, based partly on case studies.

▶ O'Connell, M. 1996. "Training as a potential profit center," *Journal of Management in Engineering* (Sept./Oct.), pp. 25-27.

Describes the focused E&T of a 260-person, five-office consulting engineering firm. As of about 1996, the firm devoted 3.5% of its annual budget to E&T. A low 5% annual turnover of personnel was one cited benefit.

▶ Stewart, T.A. 1997. *Intellectual capital: the new wealth of organizations*. New York: Currency/Doubleday.

Uses an iceberg metaphor to emphasize that, increasingly, the readily seen and measured physical and financial capital of businesses is much smaller than the less apparent intellectual capital. States that "...intellectual capital has become so vital that it's fair to say that an organization that is not managing knowledge is not paying attention to business."

Visit this website

▶ "Learnon.org" (http://learnon.org), a website maintained by the American Society for Engineering Education, provides access to over 10,000 graduate and continuing education courses in engineering. Searches can be conducted by subject matter, geographic location, cost, and course provider.

The mind is not a vessel to be filled but a fire to be kindled.

Plutarch

Improving Personal and Organizational Productivity

Success inevitably comes our way as we acquire, develop, and apply management and the leadership knowledge, skills, and attitudes suggested and described in this book. With the exhilaration of success, especially the success of doing significant work and making meaningful contributions, comes the realization that we, as individuals and organizations, could achieve much more. We are surrounded by an ever-growing galaxy of varied and exciting opportunities and possibilities. We may find ourselves saying, "so much to do, so little time."

While working harder may seem to be the answer to exploring new opportunities and pursuing new possibilities, working smarter is likely to yield better results. The lessons in this part of the book offer advice on ways to improve personal and organizational productivity so that we can accomplish more. Suggestions range from improving awareness and effective use of our subconscious mind to strengthening project management.

We Don't Make Whitewalls: Work Smarter, Not Harder

We know where most of the creativity, the innovation, the stuff that drives productivity lies—in the minds of those closest to the work.

Jack Welch

*H*ow many times have we received or given the advice to "work smarter, not harder?" Whether 10, 100, or 1,000 times, I suspect that the number of times the advice was heard was one or two orders of magnitude greater than the number of times the advice was heeded.

Nevertheless, the advice is both simple and sound. While most of us are willing to work hard and do whatever is needed to get a job done, few would quarrel with the concept of working smarter. To "work smarter" means to identify problems in a process, resolve the problems, and incorporate solutions back into the process. This approach, which, in its more formal application is commonly called reengineering (Hammer and Stanton, 1995), seeks increased efficiency while maintaining or enhancing quality. Working smarter also implies benefits to both personnel and the organization, such as putting in fewer hours to complete tasks, increasing personal time, incurring less stress, obtaining results that are more likely to meet expectations, and increasing profits.

So why do we work harder instead of smarter? I've found that there are two obstacles: First, we often lack the will or self-discipline to make the up-front investment in time and energy needed to examine the way we do things and to determine if there is a better way. Time is a precious commodity, and it is difficult to carve out sufficient time to study the way we spend our time. Second, and more troubling, we recognize—consciously or subconsciously—that if we truly examine how we do our work, we are very likely to find needed improvements. These improvements will require change, and change is difficult.

Here are two suggestions for finding ways to work smarter. The first suggestion is at the personal level. Set aside some quality time to think about one work-related process that you frequently do by yourself. Examples might be planning, conducting, and following up on a meeting of a committee you chair; writing a monthly report; calling on a potential client; or giving a talk. Search for ways to do it smarter, and consult with selected colleagues for ideas and input. An analysis of how smart we work as individuals, contrasted with how many hours we work, might reveal new ways to handle the parts of the process that require the most of our time and effort.

> *Some people actually work as little as half the time they are at work. These people create a window of opportunity for you to succeed. Don't worry about being obligated to work more hours to beat the competition. You probably don't have to. Instead, if you commit to working all the time you are at work, you will probably come out well ahead of your competition.*
>
> Bill Fitzpatrick

The second suggestion for working smarter involves an ad hoc group effort. Select some routine process that is regularly done by a group of personnel. Examples are preferably interdepartmental processes, such as assembling a proposal, designing a manufacturing process, performing a property survey, or preparing a set of plans and specifications. Gather a cross section of individuals who have a role in the process, even if it appears minor. For example, if the process is preparing a set of plans and specifications, you could include not only engineers but also clerical, CADD, and marketing personnel. Once you assemble the group, schedule a work session of predetermined duration. Establish a non-threatening atmosphere by prohibiting overly critical comments on ideas that are offered, and encourage everyone to participate on an equal basis.

Facilitate the work session by instructing the group to construct a flowchart or some other detailed description of the selected process as it currently exists. Then, brainstorm ways to collectively work smarter to create a more effective process. Narrow the ideas until workable process changes becomes evident, and discuss how they will, on balance, improve the process. Implement the changes, at least a portion of them, on a trial basis.

Group efforts such as this are successful because each member of the group stands to benefit from the improvements. There are three additional reasons:

- Most people want to contribute to a team effort.
- Non-threatening group efforts are typically very creative and synergistic.
- Individuals closest to a process are in the best position to improve it.

Mastering the concept of working smarter, not harder, requires discipline and courage to look inward—within yourself and your organization. Valuable beds of knowledge underlay your organization, but they must be "sought out, mined, and used" (O'Dell and Grayson, 1998). Knowledge management authors Carla O'Dell and C. Jackson Grayson, Jr. observe that "cave dwellers froze to death on beds of coal. Coal was right under them, but they couldn't see it, mine it or use it." In a similar fashion, the typical public or private organization is underlain by valuable beds of knowledge. And like the cave dwellers, if those organizations don't search for, mine, and use that knowledge, they may "freeze to death."

> *Don't tell me how hard you work. Tell me how much you get done.*
>
> James Ling

Suggestions for Applying Ideas

Read the following story and consider whether unexamined wasteful practices occur within your organization

An ambitious and inquisitive worker starts a new production line job at a tire manufacturing plant. After one week, he conscientiously asks why each tire is wrapped in brown paper before shipping. His supervisor's answer: "To protect the whitewalls." The new worker's response: "But we don't make whitewalls; the plant stopped doing that 10 years ago."

> *Never mistake motion for action.*
> Ernest Hemingway

▶ Are you still wrapping tires in brown paper or its equivalent?

Avoid thinking that reorganization is the only option for solving organizational problems

▶ An organization is a team, or perhaps a group of teams. The three team essentials are a strong and shared commitment to a goal, if not a vision; diversity, that is, an optimum mix of players covering all the necessary bases; and an effective operational structure.

> *I was to learn later in life that we tend to meet any situation by reorganizing; a wonderful method it can be for creating the illusion of progress while producing confusion, inefficiency and demoralization.*
> Petronius

▶ Therefore, recognize that reorganizing, or reorganizing again, addresses only the last of the three team essentials. Before reorganizing, ask if the cause of the problem, and therefore the solution to, unsatisfactory results might lie instead with lack of goals and/or commitment to them or be traced to misfit players.

▶ Individuals are often the root cause of poor performance, just as they are frequently the source of exemplary performance. In the former case, they should have an opportunity to receive critiques of their behavior and possibly change it and their performance. When the performance or behavior of one or a few individuals is deteriorating or unacceptable, there is a tendency to reorganize the affected unit or even the entire organization. This shotgun approach is sometimes taken in lieu of the rifle approach of personally confronting the problem person or personnel.

Apply some of Og Mandino's (1968) success secrets to help each of us work smarter and live wiser

▶ "Form good habits and become their slaves."

▶ Respect ourselves and, as a result, "zealously inspect" whatever may enter our bodies, minds, souls, and hearts.

> *In truth, the only difference between those who have failed, and those who have succeeded lies in the difference of their habits. Good habits are the key to all success... I will form good habits and become their slave.*
>
> Og Mandino

▶ Persist until we succeed, never allowing a day to end in failure.

▶ Appreciate our individual uniqueness, and recognize that none of us is "on this earth by chance." Each of us should search for our unique purpose.

▶ Master our emotions, recognizing that "unless my mood is right the day will be a failure."

▶ Aim high, always striving to improve, recognizing that "to surpass the deeds of others is unimportant; to surpass my own deeds is all."

▶ Act now, recognizing that dreams, plans, and goals are worthless without action.

Get past gatekeepers by trying some "work smarter" tactics

"Gatekeepers" are those sometimes overly protective administrative assistants and others who may prevent you from talking by telephone with a key person (Grugal, 2002b).

▶ Call early or late; "bosses" often start work early in the day and are among the last to leave at the end of the day.

▶ Ask the gatekeeper to help you set up a definite time for a telephone meeting as an alternative to you repeatedly calling on the chance that you will connect.

▶ Leave an intriguing message, via voicemail with the gatekeeper. I once left this voicemail message. "I have a $1,000 opportunity." The person quickly called back. However, the opportunity was a request to have his organization donate $1,000 to a professional society.

▶ Mention a third party known to the person you are trying to contact by telephone. For example, "Mary Jones suggested that I call you to discuss your engineering services needs." However, do this only if the referral is absolutely legitimate.

Adopt the seven habits of highly successful people to help you work smarter (Covey, 1990)

▶ Be proactive

▶ Begin with the end in mind

▶ Put first things first

▶ Think win-win

▶ Seek first to understand, then to be understood

▶ Synergize

▶ Sharpen the saw (i.e., renew our physical, spiritual, mental, and social/emotional dimensions)

Avoid continuing or adopting the seven habits of highly ineffective people (Green, 1995)

- ▸ Poor listening
- ▸ Negative thinking
- ▸ Disorganization
- ▸ Inappropriateness (i.e., failing to recognize that there is a time and place for everything)
- ▸ Decisions by default
- ▸ Randomization (i.e., performing tasks in random order rather than in logical sequence)
- ▸ Procrastination

Leverage the work already completed or committed to—a little investment on the margin could realize a large return on the margin

- ▸ Your firm just completed a design that provided your client with an innovative, cost-effective solution.
  - • *Possible small incremental investment:* Co-author a paper on the project with your client, and present it at a conference likely to be attended by potential clients.
  - • *Possible large incremental return:* Recognition for existing client and contracts with new clients.
- ▸ You're the "bail-out" person in your firm. Troubled projects are given to you with the request that you turn them around. While somewhat satisfying, this process is getting tiresome because it typically requires a major effort on your part.
  - • *Possible small incremental investment:* Draft a "lessons learned" memorandum and share it with project managers, perhaps in a workshop setting.
  - • *Possible large incremental return:* Reduce future problem projects and create new opportunities for you.
- ▸ Your government agency is arranging an internal seminar on effective written communication.
  - • *Possible small incremental investment:* Invite other organizations, such as consulting firms and other agencies, to attend on the premise that most people like to learn.
  - • *Possible large incremental return:* Improved relations and new contacts. While working at a university, I once helped to arrange a seminar that followed this advice. One of the invitees was a local community leader. On leaving the event, he sincerely and enthusiastically thanked me and said something like, "This is the first time I've been invited here when the university didn't want something."

▶ The "i's" have been dotted and the "t's" crossed; copies of the final report are about to be mailed to your client.

- *Possible small incremental investment:* Drive or fly to your client and hand deliver the box of reports. While there, thank your client liaison and his/her staff for their assistance during the project.
- *Possible large incremental return:* An extra measure of appreciation, on behalf of the client, for being appreciated.

▶ Consider some of your or your organization's recent accomplishments. You understandably want to move on to other projects. However, before doing so, might there be ways in which you can creatively leverage your efforts?

Read the following related lessons

▶ Lesson 3, "Smart Goals"

▶ Lesson 30, "Education and Training: From Ad Hoc to Bottom Line"

▶ Lesson 33, "Delegation: Why Put Off Until Tomorrow What Someone Else Can Do Today?"

▶ Lesson 34, "TEAM: Together Everyone Achieves More"

▶ Lesson 35, "Virtual Teams"

▶ Lesson 48, "AH HA! A Process for Effecting Change"

Study one or more of the following sources cited in this lesson

▶ Covey, S.R. 1989. *The 7 habits of highly effective people: restoring the character ethic.* New York: Simon & Schuster.

▶ Fitzpatrick, B. 1997. *100 action principles of the Shaolin.* Natick, Mass.: American Success Institute.

▶ Green, L. 1995. "The 7 habits of highly ineffective people." *American Way* (Aug. 15), pp. 56-60.

▶ Grugal, R. 2002b. "Get past the gatekeepers." *Investor's Business Daily* (Oct. 22).

▶ Hammer, M., and S.A. Stanton. 1995. *The reengineering revolution: a handbook.* New York: HarperCollins. (Provides an introduction to reengineering.)

▶ Mandino, O. 1968. *The greatest salesman in the world.* New York: Bantam Books.

▶ O'Dell, C., and C.J. Grayson, Jr. 1998. *If only we knew what we know: the transfer of internal knowledge and best practice.* New York: Free Press, p. ix

Refer to one or more of the following supplemental sources

▶ Baker, M. III. 1997. "Reengineering an engineering consultant." *Journal of Management in Engineering* (March/April), pp. 20-24.

Describes the apparently warranted and successful reorganization of an entire consulting engineering firm.

▶ Messersmith, J., and A. Schrader. 2003. "Bringing people, processes, and the work place together to create high-performance work environments." Forum, *Leadership and Management in Engineering* (Jan.), pp. 5-7.

Explains how the convergence of workforce, work flow, and workplace establish an organization's effectiveness. Examination of all three factors, and possible improvements to some or all of them, can improve financial performance, increase productivity, and reduce product or service delivery time.

Find the essence of each situation, like a logger clearing a log jam.
The pro climbs a tall tree and locates the key log, blows it, and
lets the steam do the rest. An amateur would start at the edge
of the jam and move all the logs, eventually moving the key log.
Both approaches work, but the "essence" concept saves time and effort.
Almost all problems have a "key" log if we learn to find it.

Fred Smith

The Power of Our Subconscious

A man cannot directly choose his circumstances, but he can choose his thoughts, and so indirectly, yet surely, shape his circumstances.
James Allen

Writing and designing are creative activities and, in both cases, we often "run into a brick wall." When writing, the "wall" often appears when we know what we want to say but can't find the words to say it. The design "wall" looms when we first truly understand all the constraints and expectations for a project but can't see a way to even come close to satisfying them. Oftentimes, "out of the blue," the words or the way appear. What is the source, what is this force, and how can we use it even more effectively?

Psychiatrist M. Scott Peck (1997) calls the mysterious source and force our unconscious mind. He says, "The conscious mind makes decisions and translates them into actions. The unconscious mind resides below the surface; it is the possessor of extraordinary knowledge that we aren't naturally aware of."

Motivational author Napoleon Hill devoted decades to studying successful people and has this to say about the subconscious mind (1960): "Your subconscious mind works continuously, while you are awake, and while you sleep." He also suggests that our subconscious mind can actually be directed to work for us:

> The subconscious mind will translate into its physical equivalent, by the most direct and practical method available, any order which is given to it in a state of belief, or faith that that order will be carried out.

As a Man Thinketh, by James Allen (1983), refers to the power of the mind as a "garden, which may be intelligently cultivated or allowed to run wild"; regardless of whether the mind is cultivated or neglected, it will "bring forth." Allen goes on to say that "thought-forces and mind-elements operate in the shaping" of a person's "character, circumstances, and destiny." In other words, we tend to become what we repeatedly tell ourselves we are (Singleton, 1990). So talk to yourself, but be careful what you say; it is very likely to come true!

In his book, *The Power of Your Subconscious Mind* (1963), theologian and scientist Joseph Murphy portrays the subconscious mind as being receptive, impressionable, sleepless, non-reasoning, creative, eager, action- or execution-oriented, intelligent, idea rich, and ageless. Murphy explains: "If you imagine an objective clearly, you will be provided with the necessities, in ways you know not of, through the wonder-working of your subconscious mind."

> *Worry consists of creating mental pictures of what you do not want to happen. Confidence is creating mental pictures of what you want to happen.*
>
> John C. Maxwell

What has any of this to do with engineering managing and leading? The subconscious mind is a powerful tool for working through difficult problems in our professional or personal lives. Rather than physically plug it in and forcefully apply it to the task at hand, we can relax and let it work in the background, creatively, within us. The answers we seek, the options we need to try—all will come forward.

Suggestions for Applying Ideas

Understand and appreciate the workings of the conscious and subconscious minds using these analogies and metaphors

► The conscious mind is the camera and the subconscious mind is the film (Murphy, 1963). Point your "camera" at the things you want to develop.

► The conscious mind sees reality, while the subconscious mind cannot tell the difference between reality seen by the conscious mind or that which is imagined by the conscious mind (Tice, 2002). Therefore, consciously imagine and visualize those good things you desire, and your subconscious mind will accept and work on them as though they were an evolving reality.

► The conscious mind selects and plants seeds, and the subconscious mind germinates and grows them. Select seeds in accordance with what you want to harvest (Murphy, 1963).

> *Our life is what our thoughts make it.*
>
> Marcus Aurelius

► If the conscious mind, exercising its free choice selects a worthy direction, the subconscious mind, directed by providence, directs the step (Prov. 16:9).

► The subconscious mind is an empty file folder, the conscious mind is the filer. Sift and winnow that which appears in your in basket, and file only that which you intend to act on.

▶ Your conscious mind is the cause; your subconscious mind, the effect. Choose your causes carefully (Murphy, 1963).

▶ The conscious mind is a part-time worker while the subconscious mind is a full-time worker; it never sleeps (Murphy, 1963). Use the limited time available with your conscious mind to direct and fully utilize the continuous, creative efforts of your subconscious mind.

The greatest discovery of my generation is that man can alter his life simply by altering his attitude of mind.
William James

▶ The conscious mind is the ship's captain and the subconscious mind a fast ship and with an excellent crew (Murphy, 1963).

▶ The conscious mind is the front burner, where attention is focused. The subconscious mind is the back burner using a process that "mixes, blends and simmers ingredients into a tasty meal" (Carlson, 1997). Feed the back burner of your mind with a "list of problems, facts, and variables, and possible solutions." Let them simmer and expect a pleasing result.

Use Hill's technique for realizing the potential of your subconscious mind (Hill, 1960; Singleton, 1990)

▶ Avoid idle wishing; instead use your conscious mind to specifically define in words, images, and feelings what you want to accomplish and when. For example, if you are engaged in a writing project, define the message you want to communicate and the responses you want to elicit. If you are performing a design, define the constraints and expectations. Envision the thoughts, words, and actions of the eventual uses of your design.

If one advances confidently in the direction of his dreams, and endeavors to live the life which he has imagined, he will meet with success unexpected in common hours.
Henry David Thoreau

▶ Reduce that well-defined desire to writing, which will further clarify your desire and embed it in your mind's eye.

▶ Temporarily set the project aside and have faith that your subconscious mind will go to work on it.

▶ Study, observe, network, ask, imagine, experiment, risk, persist, and be positive, not necessarily with your desire in mind, but to serve as stimuli for your subconscious mind.

▶ Occasionally, consciously revisit your project and be prepared to see new ideas emerge from the work your subconscious mind has been doing, enabling you to take steps toward achieving that which you desire.

Let your subconscious mind work while you sleep.

▶ Try using your subconscious mind during sleep as suggested by inventor Ray Kurzweil, whose creations include the Kurzweil Reading Machine for the blind "which converts ordinary books, magazines, and other printed material to speech" (Mink, 2001).

▶ Before going to sleep, think about a specific issue or problem, professional or otherwise.

▶ Don't try to resolve the issue or solve the problem. Instead, carefully define it and think of attributes of the resolution or solution.

In human affairs the willed future always prevails over the logical future.
Rene Dubos

▶ Your subconscious mind may be stimulated to work on the problem while you sleep. The dream state is not bound by the restrictions we typically place on our conscious thoughts.

▶ On awakening, immediately recall the issue or problem and look for new insights or even a resolution of the issue or solution to the problem.

Read the following related lessons

▶ Lesson 2, "Roles—Then Goals"

▶ Lesson 3, "Smart Goals"

▶ Lesson 20, "Write to Find Out What We Think"

▶ Lesson 21, "Start Writing on Day 1"

▶ Lesson 31, "We Don't Make Whitewalls: Work Smarter, Not Harder"

If the desire to get something is strong enough in a person, his whole being, conscious and unconscious, is always at work, looking for and devising means to get to the goal.
Frederick Philip Grove

Study one or more of the following sources cited in this lesson

▶ Allen, J. (1983). *As a man thinketh*. Marina Del Ray, Calif.: DeVorss & Company.

▶ Carlson, R. 1997. *Don't sweat the small stuff . . . and it's all small stuff: simple things to keep the little things from taking over your life*. New York: Hyperion.

▶ Hill, N. 1960. *Think and grow rich*. New York: Fawcett Crest.

▶ Mink, M. 2001. "Inventor Ray Kurzweil: his passion to create helped give blind people their independence." *Investors Business Daily* (July 10).

▶ Murphy, J. 1963. *The power of your subconscious mind*. Englewood Cliffs, N.J.: Prentice Hall.

▶ Peck, M.S. 1997. *The road less traveled and beyond: spiritual growth in an age of anxiety*. New York: Simon & Schuster.

▶ Prov. 16:9, "You may make your plans but God directs your actions." *The Bible* (Good News Version).

▶ Singleton, M. 1990. "Programming your subconscious mind for success." *Executive Journal* (June), pp. 8-14.

▶ Tice, L. 2002. "Winners circle network with Lou Tice." An e-newsletter from the Pacific Institute (http://www.thepacificinstitute.com), April 25.

The greatest achievement was at first and for a time a dream.
The oak sleeps in the acorn; the bird waits in the egg;
and in the highest vision of the soul a working angel stirs.
Dreams are the seedlings of realities.

James Allen

Delegation: Why Put Off Until Tomorrow What Someone Else Can Do Today?

*Good managers never put off until tomorrow
what they can get someone else to do today.*
Anonymous

Delegation, by definition, is legitimately and carefully assigning part of our tasks to someone else, and we engineers, especially the younger ones among us, typically dislike it. As part of the deal, we must give up some authority while retaining responsibility for the outcome. Delegation is different from "dumping," which is getting rid of responsibility when the going gets tough. Delegation is also not giving orders, which is maintaining authority while giving someone else responsibility.

Collectively, and individually, the reasons we "can't or won't" delegate are numerous. For example, we can't find the time for the up-front investment needed to show someone how to do what we do. Or no one could possibly "do it" as well as we do, or maybe someone could, but we have no one to delegate to. Another reason is that we lack the organizational and communication skills to explain it to others. Fear of losing knowledge-based job security is a factor for some of us. Others are wary of appearing lazy or incompetent. Also, sometimes we hesitate to delegate because we fear advancement: effective delegation may lead to rapid advancement to a more responsible position for which we do not feel prepared.

Frankly, I've heard all the reasons and rationalizations for why we "can't or won't" delegate, and I have used many myself. You may convince yourself, but you are not convincing others, especially peers and those who determine your salary, bonuses, and promotions. Why? Because the reasons to delegate far outweigh the reasons not to. Consider the following benefits of delegation:

- *Frees up experienced people* to take on new responsibilities, projects, and challenges—to do things they are better prepared for or would rather do.

- Gives other members of the organization *opportunities to learn, grow, and contribute* in new ways to the work of the organization.
- Helps individuals *learn from others*. Surprisingly, the learning sometimes flows from the delegatees to delegators.
- *Reduces task costs*, that is, it tends to push the cost of each task to the lowest level possible consistent with the required results.
- *Builds resiliency* into an organization. By spreading understanding of tools and techniques to other members of the organization, more people know how to do more things. This is analogous to the concept of "strengthening the bench" on an athletic team.

> *I not only use all the brains I have, but all I can borrow.*
> Woodrow Wilson

If you still "can't or won't" delegate, then recognize that the usual consequence of not delegating within an organization is that you are likely to be relegated to the "slow track." You probably will be labeled as "not a people person" and "not a team player," and this could stymie both managerial and technical advancement. In contrast, through delegation, you help your colleagues and organizations realize some of the previously listed benefits. You demonstrate many important management and leadership perspectives and skills, not the least of which is teamwork.

Suggestions for Applying Ideas

Follow these tips for successful delegation (Walesh, 2000; Culp and Smith, 1997)

▶ Make sure you understand your responsibility and have the authority to carry it out. Typically some project or process has been delegated to you and you, in turn, are delegating a portion of it to someone else.

▶ Explain the context for the delegated task. For example, describe the overall process or project and indicate how the delegated task fits in terms of inputs and outputs.

▶ Use written procedures, also called checklists, tips, guidelines, or best practices. Ask the delegatee to suggest improvements to the written procedures.

▶ Provide, as may be appropriate, the budget and schedule for the task. Often the budget can be expressed as hours of effort. Rather than using a single completion date, the schedule could be presented as a series of milestone dates with portions of the task being completed by each milestone.

▶ Prescribe and illustrate expected outcomes and deliverables. Use factors such as accuracy, format, size, and client satisfaction.

▶ Provide or arrange for the necessary tools and resources.

▶ Do not overprescribe "how." Avoid giving orders and micromanaging.

▶ Protect the delegatee from outside intrusions, well-intentioned or otherwise.

▶ Recognize these three possible outcomes of delegation, arranged from favorable to unfavorable, and respond accordingly:

- The work is delivered as needed.
- The work will not be completed as needed but the delegator is so advised by the delegatee well before the deadline.
- The work is not going to be completed as expected, and the delegator learns about the deficiency at or after the time the work was to be completed.

> *A real leader does as much dog work for his people as he can: he can do it, or see a way to do without it, ten times as fast. And he delegates as many important matters as he can because that creates a climate in which people grow.*
>
> Robert Townsend

▶ When the outcome meets or exceeds expectations, say "thank you" in a way that clearly communicates what you appreciate and why. As noted by fiction writer and humorist Mark Twain, "I can live for two months on a good compliment."

▶ If the outcome is unsatisfactory, take action. Critique the work, not the person; avoid negative "you" messages. Hold the delegatee accountable for meeting his or her responsibility. Look for ways that the delegatee can partly or completely correct the unsatisfactory outcome.

Test yourself for delegation effectiveness by answering the following questions "yes" or "no"

▶ Are you working longer hours than almost everyone else?

▶ Do you regularly take work home?

▶ Are you frequently rushing to meet deadlines?

▶ Are top priority action items needed to fill your desired roles and achieve your established goals on the "back burner" or, worse yet, not even started?

▶ Are you the only person capable of managing the next big assignment or project?

▶ Is your staff listless and stagnant and/or do you experience high turnover?

If most or all of your answers are "yes," you are fooling and shortchanging yourself, others, and your organization (Culp and Smith, 1997).

Take your delegation effectiveness up to a new level by following these suggestions

▶ Think of a task that "only you can do." Plan to delegate it, or part of it, to someone else.

▶ Plan to delegate a task to someone to whom you have never delegated.

▶ Try one more time to delegate a task to someone who failed in the past. Maybe you were part of the problem.

▶ Urge one of your non-delegating supervisees to try delegation.

> *Identity requires responsibility because without responsibility there is no self respect. You do not know whether you could handle anything, deliver any result or take care of anyone else.*
>
> Charles Handy

▶ Identify someone from whom you will no longer accept substandard performance. Hold them accountable to meet their responsibilities the next time they come up short.

▶ Think of someone who always comes through when tasks are delegated—who does it so well that you hardly notice. Say "thank you," recalling the earlier advice to be very specific.

Read the following related lessons

▶ Lesson 29, "More Coaching, Less Osmosis"

▶ Lesson 31, "We Don't Make Whitewalls: Work Smarter, Not Harder"

▶ Lesson 45, "Our Most Important Asset"

Study one or more of the following sources cited in this lesson

▶ Culp, G., and A. Smith. 1997. "Six steps to effective delegation." *Journal of Management in Engineering* (Jan./Feb.), pp. 30-31.

▶ Walesh, S.G. 2000. *Engineering your future: the non-technical side of professional practice in engineering and other technical fields,* 2nd Ed. Reston, Va.: ASCE Press. Chapter 4, "Management of Relationships."

> *Irrevocable commitments that offer no loopholes, no bailout provisions, and no parachute clauses will extract incredible productivity and performance.*
>
> Robert A. Schuller

TEAM: Together Everyone Achieves More

One hand cannot applaud alone.
Arabian proverb

*E*arly on in my college days, my newfound friends and I—all of us equally naive—joined the same fraternity. We soon discovered that our choice was not a good one, as the fraternity had not distinguished itself when measured by indicators such as academics, athletics, and campus leadership.

Within our first year of membership, and largely by coincidence, a musically inclined two-year member of the fraternity presented a **vision:** to begin to rebuild our group's role and reputation on campus. All we had to do was win the fraternity portion of the next annual Greek Songfest. In retrospect, the significance of a campus Songfest seems trivial. But at that time and in that place, the Greek Songfest was the big event, and winning it was the big prize.

Amazingly, the vision was contagious. It spread throughout our fraternity, with some of us beginning to think that we actually might be able to achieve it. Had we shared the vision with other Greeks and the campus community, it would have been met with derision, given our group's low status. However, we began to identify with the vision and increasingly that was all that mattered.

Commitment to the vision led to individual singing auditions. Strength in *diversity* emerged. We had a good mix of basses, baritones, tenors, altos, and sopranos. The vision's author proved to be a very capable arranger and director, and other fraternity brothers filled organizational roles. A few members had no singing talent but compensated by being excellent at mouthing the words.

An organizational and operational *structure* soon emerged. Its foundation was a rigid schedule of frequent practices. A monetary fine system was instituted for missed sessions. Peer pressure increased. We were in this together; everyone was needed and held accountable.

Even though 40 years have passed, the intense Songfest effort was one of my most enlightening team experiences. It also was one of the most satisfying, because we won! We also won the next two years. By the time my class graduated, and as a result of the catalytic effect of the first Songfest win, our fraternity was one of the most highly regarded on campus, excelling in academics and leadership.

While the college musical competition may be of only nostalgic note, the team lesson learned from winning the Songfest has been of lifelong value to me. What are the elements of that useful team lesson? Before offering my answer, briefly consider another, much more recent personal experience.

A high-level task committee (TC) was formed by a national professional organization, and I was asked to be a member. The charge was to explore a potential historic change in the education of certain professionals. After some false starts and initial uncertainty, the TC soon articulated and committed to an ambitious *vision*. Using gifts and talents represented by the TC's *diversity*, definitive strategies and tactics needed to achieve the vision were developed, convincing presentations were made, and a detailed and comprehensive report was developed. The TC's highly communicative and trusting *structure* included a chairperson, support staff assistance, in-person and teleconference meetings, and numerous interactions with stakeholders. Result: the TC's recommendations were unanimously accepted by the organization's board and are now being implemented.

Again, what teamwork lesson can we learn from successful group experiences like the two cited here? I am convinced that the three key elements of successful teamwork are:

- A strong and shared commitment to an ambitious *vision*, and the bolder the better. The vision should initially appear highly desirable but unachievable. Architect and city planner Daniel Burnham suggested the power of vision-driven teamwork when he said, "Make no little plans, they have no magic to stir men's blood. Make big plans, aim high in hope and work and let your watchword be order and your beacon beauty."

> *Great discoveries and improvements invariably involve the co-operation of many minds. I may be given credit for having blazed the trail but when I look at the subsequent developments I feel the credit is due to others rather than to myself.*
>
> Alexander Graham Bell

- *Diversity*—an optimum mix of players. All the necessary bases must be covered. Factors to consider in forming a team are to seek individuals who, besides sharing the vision, collectively bring the necessary knowledge, skills, connections, and time availability.

- An effective operational *structure*. Trust and open, ongoing, intrateam communication are essential.

An exciting vision without diverse players is a dream. A talented team toiling in a vision void is poor stewardship. Engineer and educator Arthur E. Morgan said "Lack of something to feel important about is almost the greatest tragedy a man may have" (Lueba, 1971). A superb organizational structure without talented players degenerates into bureaucracy. All three elements are needed: *vision*, *diversity*, and *structure*.

Suggestions for Applying Ideas

Perform reality checks

▶ Have you, in your work, community, athletics, family, or other endeavors, recently had an uplifting team experience?
  • If not, seek out team opportunities so you don't miss out on one of life's most rewarding experiences.

▶ Can you, as an executive or principal in your organization, point to outstanding team efforts within your organization?
  • If not, your organization is probably falling far short of its potential. *Suggestion:* Articulate an ambitious vision, assemble a diverse team, ask them to achieve the vision, provide support, and get out of the way.

Appreciate that teams, especially those with ambitious expectations and highly motivated players, often need to progressively work through the four stages of team development (Brown, 1992; Martin and Tate, 1999)

▶ *Forming:* Politeness, inquiry, waiting to see what will happen, no or very little productivity.

▶ *Storming:* Disagreement, confusion, conflict, factions, some productivity.
  • The storming stage can be reduced in intensity or avoided, partly by providing opportunities for individuals to become acquainted, providing history and other context for the team's task or charge, developing meeting and communication norms, agreeing on basic terminology, encouraging individuals to share concerns, and defining possible roles of team members.

▶ *Norming:* Conflict resolution, goal setting, decision making, establishment of protocols, ownership, accountability, moderate productivity.

▶ *Performing:* Teamwork, adjustments, deliverables, full ownership and accountability, can-and-will attitude, high productivity, satisfaction, celebration.

Recognize and temper potential "red flags" in team members' personalities (Thompson, 1996)

▶ Cynical attitude

▶ Aloofness

▶ Strong need to win

▶ Preference for clearly defined short-term goals and discomfort with ambiguity

▶ Impatience with indecisiveness

▶ Irritability and defensiveness in response to criticism

> *Engineers don't always make the best team players.*
> John W. Thompson

> ► Meticulousness and intolerance of others' mistakes

> ► Excessive immersion in details

Read the following related lessons

> ► Lesson 35, "Virtual Teams"

> ► Lessons 39 through 41 in Part 5, "Meetings"

Study one or more of the following sources cited in this lesson

> ► Brown, T.L. 1992. "Teams can work great." *Industry Week* (Feb. 17), p. 18.

> ► Leuba, C.J. 1971. *A road to creativity—Arthur Morgan—engineer, educator, administrator.* North Quincy, Mass.: Christopher Publishing House

> ► Martin, P., and K. Tate. 1999. "Climbing to performance." *PM Network* (June), p. 14.

> ► Thompson, J.W. 1996. "Engineers don't always make the best team players." *Electronic Engineering Times* (Sept. 30), p. 124.

Refer to the following supplemental sources

> ► Hensey, M. 1995. *Continuous excellence: building effective organizations.* Reston, Va.: ASCE Press. Chapter 19, "Managers Have to Be Facilitators."
> Recognizes the importance of facilitation in the lives of some teams and provides a "bare minimum" list of facilitator skills. Outlines these four group or team problem solving tools: brainstorming, displayed thinking, Pareto analysis, and flow-charting.

> ► Shonk, J.H. 1997. *Team-based organizations: developing a successful team environment.* Chicago, Ill.: Irwin Professional Publishing.
> Builds on the premise that teams offer "an effective way to coordinate across organizational boundaries in solving problems and gaining employee commitment." Intended for managers and leaders, this book explains how to plan for and implement transitioning an organization from a traditional hierarchical and functional structure to more team-based organization.

Visit one or more of these websites

> ► "About Teambuilding, Inc." (http://www.teambuildinginc.com) is offered by a participatory management consulting firm led by civil engineer Peter Grazier. Many free, team-related articles are offered. Another useful feature is a discussion section where visitors can post questions for response by Grazier and others.

> ► "Team Management Systems" (http://www.tms.com.au) is maintained by Margerison-McCann, a firm that focuses on "why some individuals, teams

and organizations perform, work effectively and achieve their objectives, while others fail." Useful free features include many case studies and articles.

The society which scorns excellence in plumbing as a humble activity and tolerates shoddiness in philosophy because it is an exalted activity will have neither good plumbing nor good philosophy: neither its pipes nor its theories will hold water.

John W. Gardner

35

Virtual Teams

> *The diplomatic art of managing ad hoc partnerships and alliances will become a key executive skill.*
>
> Economist

Virtual teams allow smaller organizations to "play in the big leagues." Small organizations complement their core competencies with virtual teams, which enable them to assemble the expertise needed to compete for and, if selected, complete big projects or to take on challenging causes.

A virtual team is like a traditional team in that it has the three team essentials of common vision or purpose, necessary diversity of expertise, and effective working structure. In a traditional team, the team members are typically all employees of one organization. However, in a virtual team, the members typically are a carefully selected, eclectic mix of sole proprietors and employees of various firms who are usually linked electronically. The explosion in the use of low-cost electronic communication devices—e-mail, websites, wireless telephones, pagers, and facsimile machines—has facilitated the virtual team phenomenon. An article in the *Economist* (1999) draws this conclusion about the business impact of the Internet and other newer technologies:

> The boundaries of companies will also change. . . . Companies will find it easier to outsource and to use communications to develop deeper relations with suppliers, distributors and many others who might once have been vertically integrated into the firm.

The legal and organizational structures of the members of a virtual team or organization are secondary. Of prime concern is the extent to which members of the virtual group are able to function in service of the client or some cause. Trust is essential; it is the glue that binds the virtual team or organization (King, 1996).

Virtual teams are increasingly formed to meet the specific and demanding needs of certain clients or their projects. Participants recognize that, while any one existing organization may not be able to marshal all the necessary talents, a specialty tailored virtual team can do so.

The client or other stakeholder may or may not be aware of the distinction between being served by a traditional, in-organization project team versus a virtual project team or even a virtual organization. If the client or other entity shows interest, they have a right to know how they are being served. However, in most cases, the test of the delivery system will not be "how they are organized." Instead, the test will be "how do they perform" in meeting requirements.

Once the purpose is accomplished, the team is likely to disband, never to be re-assembled in exactly the same manner. Some team members may have worked together on virtual teams in the past and others may work together in the future—but only if driven by client or project requirements and by recent compatible, individual performance. Unlike actual teams, virtual teams do not have to "carry" anyone because they "need the work" or are on the payroll. There need be no force fits, only perfect fits.

Suggestions for Applying Ideas

Build an effective, functioning virtual team by incorporating these suggestions (Ashton and Ashton, 1999)

▶ "Work only with people you know and trust." Test the waters with a small project.

▶ Clearly define—upfront, in writing, and for each team member—the scope of services, schedule, deliverables, and budgets. In other words, prepare a project work plan and obtain buy-in from all members of the virtual team.

The way a team plays as a whole determines its success. You may have the greatest bunch of individual stars in the world, but if they don't play together, the club won't be worth a dime.

Babe Ruth

▶ Establish a communication protocol. For example, select the primary mode of communication from among the many options available, including telephone, fax, e-mail, pagers, and websites. With e-mail and website communication, costs are independent of distance, unlike telephone and fax communication. The selection of members for a virtual team and the ease and cost of communication among them should not be hindered by distances among them. E-mail and website communication supports this principle.

Bring a new virtual team together for a face-to-face meeting near the beginning of their work

▶ Do so even if a team could, from a technical perspective, conduct all of its work via electronic communication.

▶ As sophisticated and cost effective as electronic communication may be, experience strongly suggests that early minimal, but carefully planned and orchestrated, face-to-face interaction will enhance communication and help to build trust.

▶ Provide ample opportunity during face-to-face meetings for casual side conversations.

Extend the virtual team concepts and applications to volunteer efforts

▶ As described to this point, "virtual team" refers to the employment environment.

▶ Many of us also serve as volunteers on ad hoc committees, task forces, and other groups formed by professional societies, community groups, and service clubs.

▶ Instead of seeing these temporary groups as traditional committees or task forces, view them and have them function as virtual teams.

▶ Stress the team's vision or goal, celebrate and enhance the diversity of expertise, and establish an effective structure for communicating and otherwise working together.

▶ Adopt a sense of urgency, similar to that which is likely to prevail in the employment environment.

▶ If feasible, arrange a face-to-face meeting soon after forming the team.

▶ Then get on with realizing the goal or creating the plan to achieve the vision.

Read the following related lessons

▶ Lesson 18, "Balance High Tech and High Touch"

▶ Lesson 34, "TEAM: Together Everyone Achieves More"

▶ Lessons 39 through 41 in Part 5, "Meetings"

Study one or more of the following sources cited in the lesson

▶ Ashton, A., and R. Ashton. 1999. "Long distance relationships." *Home Office Computing* (July), pp. 52-55.

▶ *Economist.* 1999. "When companies connect" (June 26), pp. 19-20.

▶ King, R.T. 1996. "The company we don't keep." *The Wall Street Journal* (Nov. 18).

Refer to the following supplemental source

▶ Peters, T. 1994. *The pursuit of WOW!* New York: Vintage Books. Chapter entitled "Tomorrow's Strange Enterprises."

Uses a Hollywood film production and a Carnegie Hall musical production as examples of virtual teams. Notes that leaders of virtual teams tend to recruit key members rather than have them assigned, increasing the likelihood of compatibility.

Visit one or more of these websites

▶ "Virtual Projects" (http://www.vrtprj.com) is maintained by Rainer Volz, an IT and project management consultant. Features include links to mostly European project management and virtual organization websites and short reviews of project management and virtual team books.

▶ "Livelink Virtualteams" (htttp://www.virtualteams.com/index.asp) is maintained by NetAge. Provides many free virtual team articles.

To own resources is a mistake. Instead, you need instant access to the best resources from wherever, whenever, to get the job done. . . . Now impermanence and improvisation are markers for success.
Tom Peters

Fruits of Effective Project Management

Project Management: The application of knowledge, skills, tools and techniques to project activities in order to meet or exceed stakeholder needs and expectations from a project.

Project Management Institute

*I*f project management is the tree, then its fruit is the successful delivery of required products and services, on time and within budget, to the satisfaction of the client and client stakeholders as well as the project organization or team. Carefully cultivated and cared for, the project management tree will flourish, bear much fruit, and provide a continuing harvest for an organization and its members. In contrast, if not cared for, the fruits will be diminished in quality and reduced in number.

Project management is the set of processes by which public and private organizations marshal resources to design, develop, implement, and maintain required products and services. The manner in which an organization manages its projects is crucial to satisfying stakeholders and, in the case of a business, developing clients and being profitable. Whether they are large, small, sophisticated, or basic, all projects require careful project management.

The word "project" is used broadly in this lesson. It is not necessarily limited to technical work or contracts with clients. For example, some projects are for external stakeholders, such as preparing a report or designing a structure; some are for internal stakeholders, such as preparing a business plan or conducting a workshop for employees. Projects are distinguished from processes in that processes are ongoing and repeated, but projects tend to be unique. The Project Management Institute defines a project as follows (2000):

> A temporary endeavor undertaken to create a unique product or service. Temporary means that every project has a definite beginning and a definite end. Unique means that the product or service is different in some distinguishing way from all similar products or services.

Essentially everyone in an organization, business or otherwise, is at least indirectly involved in projects. Each person can contribute to the successful completion of projects and derive satisfaction from the team's and organization's achievements.

Successful project management propagates throughout an organization, in many and varied positive ways, thus significantly contributing to an organization's success. Conversely, mediocre or failed project management has widespread, negative, and sometimes devastating impacts on an organization. Accordingly, project management should be a major, if not the principal, focus of an organization's energies. Like the trunk of a tree, it is the organization's supporting structure.

There are nine "fruits" of effective project management (Walesh, 1996):

1. *Happy, satisfied clients, customers, and constituents.* Effective project management means quality, that is, meeting requirements. Project management is the activity closest to an organization's clients, customers, and constituents; they immediately and continuously receive the results, positive or negative, of the way projects are managed.

2. *Profitability.* I have a friend who once said, tongue in cheek, "We lose a little on each project, but we make it up in volume." Clearly, a business needs profit to survive, let alone thrive, and projects are the profit (or loss) generators.

3. *Reduced likelihood of client dissatisfaction, disputes, and liability exposure.* The majority of claims against consulting firms come directly from their clients, not from third parties; and they are probably traced to failed project management.

4. A fantastic on-the-job arena for *teaching and learning.* There is no "make-believe" here. Individual or corporate technical and non-technical capabilities can be markedly enhanced during the conduct of projects. Experience—good and bad—of senior personnel can be readily shared with junior personnel within the project management process.

5. The setting in which existing technical and non-technical *methods can be improved and new approaches developed.* The idea that "necessity is the mother of invention" is clearly demonstrated in effective project execution.

6. A forum in which individuals and the organization can identify *future needs* of their clients and stakeholders. Following through and meeting those needs enhances an organization's ability to serve.

7. *New projects* with existing clients, customers, and constituents because they are pleased with the results of recent and current projects and because project team members learn of additional client plans and needs. Effective project management earns the trust of the individuals and organizations being served, which is essential to earning the privilege of serving them again.

8. Satisfied clients become the source of ideas, experience, and references, earning an organization the opportunity to provide similar services to *new clients.*

9. *Personal satisfaction* through team achievement.

As logical as all the preceding seems to me, and possibly to you, I marvel at the management—or, more specifically, the lack of it—of projects in both public and private organizations. Proven principles and sound steps are routinely ignored. Are you and your organization harvesting your share of the fruits of effective project management? Or are you losing a little on each project, with the hope of making it up in volume?

Suggestions for Applying Ideas

Focus constructively on the three demands of project management

▶ Most projects encounter these three potentially conflicting demands. Satisfying two of the three requirements is easy, but reconciling all three is difficult.
  • Providing deliverables that satisfy client and other stakeholder requirements.
  • Meeting the schedule.
  • Staying within the budget.

"Look before you leap" from being a member of a project team (project engineer/scientist) to being a project manager

▶ Consider the following important—but different—characteristics of both jobs. Make sure you either have, or are developing, the necessary knowledge, skills, and attitudes to be a successful project manager (Walesh, 1996; KLO People Dynamics, 2002; Thornberry, 1989).

Project Engineer, Scientist, etc.	Project Manager
Focuses on details	Plans and monitors the big picture
Does it all	Delegates
Works mostly with things	Works mostly with people
Motivates self	Motivates others
Wants to do it right	Wants to do right by the client
Produces	Manages and leads
Derives satisfaction from technical tasks	Derives satisfaction from group accomplishments
Can "get by" as communicator	Speaks and writes well
Seeks logic and order	Handles ambiguity
Works on projects brought in by others	Brings in projects
Sees many role models	Sees few role models

Arrange for one member of your team, department, or organization to join the Project Management Institute (PMI) on a one-year trial basis

Project managers fall into three basic categories: those who watch things happen, those who make things happen, and those who wonder what happened.

Sunny and Kim Baker

► Review PMI's monthly magazine, *PM Network*; its monthly journal, *Project Management Journal*; and the organization's newsletter and selected other publications.

► Glean and share potentially useful concepts, information, and methodologies in the interest of improving project management. Possibly rotate the reviewing and sharing responsibility among personnel on a monthly basis.

► The overall thrust is to plug into, at least on an experimental basis, the premier project management professional organization.

Recognize the difference between "project" and "process"

► While "project" is used broadly in this lesson, and includes but goes beyond contracted services, not everything we do in our organization is part of a project (Martin and Tate, 2001a):

- "There are only two ways work gets done: through . . . processes or through projects."
- "Everything new or improved that happens requires a project. All ongoing operations require . . . processes."

► This table lists the essentials of projects and processes:

Project	*Process*
Temporary—has a beginning and an end	Ongoing—the same process is repeated over and over again
Produces a unique output or deliverable	Produces the same output each time the process is run
Has no predefined work assignments	Has predefined work assignments

► When a process is to be significantly improved or redesigned, that effort becomes a project.

► Distinguish between projects and processes and manage accordingly.

Read the following related lessons

► Lesson 34, "TEAM: Together Everyone Achieves More"

► Lesson 35, "Virtual Teams"

► Lesson 37, "Every Project Is Done Twice"

Study one or more of the following sources cited in this lesson

▶ KLO People Dynamics. 2002. "How to avoid the Peter Principle when selecting managers." http://pages.prodigy.net/klo_people_dynamics/pp01acec.htm.

▶ Martin, P.K., and K. Tate. 2001a. "Not everything is a project." *PM Network* (May), p. 25.

▶ Project Management Institute. 2000. "A guide to the project management body of knowledge." Newtown Square, Pa.: PMI.

▶ Thornberry, N.E. 1989. "Transforming the engineer into a manager: avoiding the Peter Principle." *Civil Engineering Practice* (Fall), pp. 69-74.

▶ Walesh, S.G. 1996. "It's project management, stupid!" *Journal of Management in Engineering* (Jan./Feb.), pp. 14-17.

Refer to one or more of the following supplemental sources

▶ Baker, S., and K. Baker. 1998. *The complete idiot's guide to project management.* New York: Alpha Books.

Adopts a broad definition of project, consistent with this lesson, and addresses an audience that extends way beyond engineers. Uses a breezy, sometimes humorous style to offer numerous project management tips. Concludes with a useful glossary. A stimulating counterpoint for more traditional project management books.

▶ DPIC Companies. 1993. *The contract guide: DPIC's risk management handbook for architects and engineers.* DPIC.

Provides detailed useful contract/agreement language with commentaries.

▶ Hayden, W.M., Jr. 2002. "Navigating the white water of project management." *Leadership and Management in Engineering* (April), pp. 20-22.

Stresses the importance of communication in effective project management stating that "the greatest adversaries to project success are silence and undecipherable messages." Notes the inevitability of conflict as a project proceeds and states that "conflict may be perceived as either a danger to be avoided or an opportunity to be embraced."

▶ Spinner, M.P. 1997. *Project management principles and practices.* Upper Saddle River, N.J.: Prentice Hall.

Targets engineers and is "intended to train the readers, in basic project management principles for directing the course of a project." Focuses on planning and scheduling the timing and costs of a project, controlling timing and costs, and allocating personnel effectively during project execution.

▶ Walesh, S.G. 2000. *Engineering your future: the non-technical side of professional practice in engineering and other technical fields,* 2nd Ed. Reston, Va.: ASCE Press.

Chapter 5, "Project Management," offers an introduction to project management. Topics discussed include the centrality of project management, project time and task management, project planning, project monitoring and control, and project post-mortems.

Subscribe to this e-newsletter

▶ "Point Lookout," a free weekly e-newsletter from Chaco Canyon Consulting. Featured are essays and white papers on teamwork, conflict, and project management. Many previously published essays and white papers are available at no cost. To subscribe, go to http://www.chacocanyon.com/.

Visit one or more of these websites

▶ "Michael Greer's Project Management Resources" (http://www.michael-greer.com/) is maintained by consultant Greer, whose business offers project management education and training, consulting, and products. Offered free are a project management bibliography and many articles including "10 Guaranteed Ways to Screw Up Any Project" and "Project 'Post-mortem' Review Questions."

▶ "The Project Management Institute" (http://www.pmi.org/) is the official website of the 100,000-member PMI. Included are membership information, conference and seminar announcements, calls for papers, and a bookstore.

The road to success is not doing one thing 100 percent better but doing 100 things one percent better.

H. Jackson Brown, Jr.

37

Every Project Is Done Twice

The beginning is the most important part of the work.

Plato

Organizations that are not satisfied with their project managers' performance take a variety of actions: threaten, offer rewards, reduce the number of project managers, hire or appoint new project managers, provide education and training for and/or certification, prepare project management handbooks, purchase project management software, and convene gatherings of project managers at which ideas and information are exchanged.

All things are created twice. There is a mental or first creation, and a physical or second creation.

Stephen Covey

These and other tactics, if carefully selected and optimally combined, can be useful. However, in my view, the most fundamental and powerful "secret" of effective project management is the title of this lesson. Unless project managers are convinced that every project is done twice—and twice in a positive sense—all the threats, treats, training, and tools will be of little use.

There is a smart way to do a project twice, and there is a not-so-smart way. The not-so-smart way is wasteful: the first time, the project, or major portions of it, is done wrong because of poor or no planning. Haphazard, high-profile, initial activity is viewed as progress. Eventually much of the work has to be redone, that is, done a second time. Negative outcomes typically include waste, frustration, loss of clients or constituents, and, for businesses, low or no profitability.

The smart way to do a project twice is to first *mentally* create the entire project and then, and only then, *actually* create—that is, execute—the project. In other words, plan the work and then work the plan. Think through, then do. This approach requires a high degree of self and organizational discipline. However, it yields a greatly enhanced probability that the second and ultimate creation will be achieved on time and within budget and will meet client and other project requirements.

The book *Zen and the Art of Motorcycle Maintenance,* by quality thinker Robert Pirsig (1974), ostensibly describes a process to use for a motorcycle repair project. But the advice is applicable to the many projects underway in our public and private organizations. Before beginning the project, each of us is urged to do the following:

> *Thought is behavior in rehearsal.*
> Sigmund Freud

List everything you're going to do on little slips of paper, which you then organize into proper sequence. You discover that you organize and then reorganize the sequence again and again as more and more ideas come to you. The time spent that way usually more than pays for itself in time saved on the machine and prevents you from doing fidgety things that create problems later on.

On considering the theme of this lesson, you may be inclined to dismiss it as simple or obvious. Perhaps it is. However, if asked, could you immediately show the written project plan for all of the contracts or other projects on which you are the manager or a team member? If project plans exist, have all team members bought into them, and are the plans current?

> *I'll give you a six-word formula for success: think things through— then follow through.*
> Eddie Rickenbacker

If your answer is "yes" to these questions, you are a rare individual or in a rare organization. And you and your organization are probably harvesting more than your share of project management fruits. If your answer is "no," if you are dissatisfied with your project management yield, maybe a fundamentally new approach is needed. Otherwise, if you do what you did, you'll get what you got.

Suggestions for Applying Ideas

Include some or all of the following in your project plan

▶ *Scope of services.* Possibly include or reference a readily available copy of the scope section of your contract or agreement. Address identification and resolution of scope creep.

▶ *Team directory.* List and provide contact information for key project staff within your organization, as well as the client's and possibly other stakeholders' organizations.

▶ *Communication.* Who are the lead contacts and how and when will they communicate?

▶ *Work task breakdown.* Identify tasks, related hours and other resources, and responsible individuals.

▶ *Deliverables and schedule.* Explain what is to be delivered (e.g., report, plan set, cost estimate, meetings) and when.

▶ *Budget.* Provide budget (labor and expenses) for each task/individual/ department.

▶ *Written guidelines, checklists, tips, and best practices.* Refer to these documents to avoid reinventing the wheel.

▶ *Documentation and filing.* Prescribe how meetings, analyses, design, and other project tasks will be documented and filed for use during and possibly after the project.

> *P6: Prior proper planning prevents poor performance.*
> Anonymous

▶ *Billing procedure.* Describe procedure and format for and frequency of invoices. Consider including a short status report with each invoice.

▶ *Monitoring.* Explain how deliverables, budget, and schedule will be monitored and, if needed, modified.

Try this "low tech" risk assessment method as part of your next project plan (Martin and Tate, 2001b)

▶ Assemble the team that will conduct the project.

▶ Give each one sticky notes.

▶ Invite team members to brainstorm risks associated with the project, write them on the sticky notes, and place them anywhere on a newsprint pad, white board, or other readily visible area.

▶ Possible potential problems are "client likes many meetings," "subsurface conditions are problematic," and "high-quality aluminum is difficult to obtain."

> *If your project doesn't work, look for the part that you didn't think was important.*
> Arthur Bloch

▶ Eliminate duplicates and clarify the meaning of risks as needed.

▶ Draw X and Y axes on the newsprint pad, white board, or other readily visible area. Write "low, medium, and high probability" on the Y axis. Write "low, medium, and high impact" on the X axis.

▶ Using group consensus, place the sticky notes on the axes based on their probability and impact.

▶ Focus on the high probability–high impact quadrant and brainstorm mitigation measures.

▶ Include selected measures in the project plan.

Watch for symptoms of the "project plan avoidance syndrome" in yourself and others on the team (Cori, 1989)

▶ Solving problems as they arise is more satisfying.

Why do we never have enough time to do it right but always enough time to do it over?

Anonymous

▶ Neither time nor environment are available for the focused thinking.

▶ Necessary labor is not in the budget.

▶ Written project work plans provide a means to monitor and measure project manager assessment and we fear that evaluation.

Read the following related lessons

▶ Lesson 34, "TEAM: Together Everyone Achieves More"

▶ Lesson 35, "Virtual Teams"

▶ Lesson 36, "Fruits of Effective Project Management"

Study one or more of the following sources cited in this lesson

▶ Cori, K.A. 1989. "Project work plan development." Paper presented at the Project Management Institute and Symposium, Atlanta, Ga. (Oct.).

▶ Martin, P.K., and K. Tate. 2001b. "A project management genie appears." *PM Network* (Aug.), p. 18.

▶ Pirsig, R.M. 1974. *Zen and the art of motorcycle maintenance.* New York: Bantam Books, p. 284.

Read the following supplemental source

▶ Rad, P.F. 2001. "From the editor." *Project Management Journal* (June), p. 3.
Defines risk management as "the systematic process of identifying, analyzing and responding to a project's unplanned events" and goes on to describe three categories of risks.

The secret of getting ahead is getting started.
The secret of getting started is breaking your complex overwhelming tasks into small manageable tasks, and then starting on the first one.

Mark Twain

38

Limiting Liability

> ### *We made too many wrong mistakes.*
> *Yogi Berra*

"*L*iability" is being bound or obligated according to law or equity. Liable individuals or organizations have to compensate by making a payment and/or taking certain actions. Engineers and other technical professionals incur liability, either individually or for their employer, primarily by negligence. "Negligence" means failing to exercise care and provide expertise in accordance with the profession's standard of care. For example, making a significant calculation error is likely to be considered negligence. Being honest and well-intentioned, although admirable, is not sufficient to avoid negligence. You and your organization must practice discipline to minimize the likelihood of negligence.

Besides negligence, individual engineers and their organizations can incur liability by "breach" or "fraud." Breach is "violation of a right, a duty or a law, either by an act of commission or by non-fulfillment of an obligation" (Dunham, Young, and Bockrath, 1979). An example of breach is failure to deliver plans and specifications by the date specified in a contract. Intent is irrelevant in determining breach. Fraud is an "intentionally deceitful practice aimed at depriving another person of his rights or doing him injury in some respect" (Dunham, Young, and Bockrath, 1979). An example of fraud is falsely telling a client that a regulatory agency approves a project. Intent is significant in determining fraud.

Negligence is the dominant cause of liability determinations in engineering, while breach and fraud account for a very small fraction. The smart engineer guards against negligence. How, then, can individuals help to reduce the probability of incurring liability for themselves and/or their employers? Some suggestions drawn from a much longer list (Walesh, 2000) and aimed primarily at individuals follow:

▶ *Maintain competence.* Services that fall below the "appropriate standard of care" may be deemed negligent. Your and your employer's best interests require that you remain current in the technical areas required in your practice.

▶ *Document everything.* Everything includes, but is not limited to, meetings, telephone calls, e-mails, field reconnaissance, and conversations. Assume that all of your work will someday be laid bare and examined by peers or, even worse, by opponents in litigation. While stressful, litigation involvement early in one's career can provide a memorable lesson in the importance of documentation.

▶ *Do it right the first time.* Rushing to redo work, late in a project's life, increases the probability of causing other errors and, as a result, incurring liability via negligence.

▶ *Use prudent language in contracts and agreements.* For example, don't say, "All existing information will be assembled"; instead say, "Readily available, pertinent information will be gathered." Avoid using words like all, insure, ensure, and assure.

▶ *Communicate-communicate-communicate.* Make sure that project requirements are understood by members of the project team at the outset and throughout the project. Proactively address potential problems within the project team or among the team, the client, regulators, or other stakeholders.

> *The search for someone to blame is always successful.*
> Robert Half

English philosopher, mathematician and linguist Thomas Hobbes said, "Prudence is a presumption of the future, contracted from the experience of time past." Experience informs us that negligence is the principal cause of liability in engineering. Experience also teaches that prudence achieved with individual and organizational discipline will minimize the probability of negligence and, therefore, liability.

Suggestions for Applying Ideas

Learn from failures, catastrophic and otherwise; become a student of failures; consider the failures described below and lessons learned from them

- Collapse, during construction, of the Quebec Bridge over the St. Lawrence River in 1907, killing 75 people and injuring 11. *Cause:* Failure to perform additional calculations to confirm structural integrity of increased span (Heery, 2001).
- Destruction, in 1940, of most of the Tacoma Narrows Bridge in Washington State with, fortunately, no lives lost. *Cause:* Failure to account for small stiffness relative to other suspension bridges (Delatte, 1997).

- Failure, in 1981, of walkways inside the Kansas City, Missouri, Hyatt Hotel, killing 114 people and injuring 200. *Cause:* Did not recognize increase in static loads on connections resulting from design change made during construction (Walesh, 2000; Heery, 2001).
- Slow collapse of a supermarket roof in Vancouver, British Columbia, in 1988, injuring 24 people. *Cause:* Failure to verify strength of beam and to consider consequences of removing two columns to increase retail space (*ENR*, 1988a).
- Failure, during placement, of a precast reinforced bridge segment in Aschaffenburg, Germany, in 1988, killing one construction worker and injuring five others. *Cause:* Inadequate design of temporary cable system used to move bridge segment from a barge to the bridge piers (*ENR*, 1988b, 1990b).
- Collapse of a school cafeteria wall in New York City during a storm, killing nine children. *Cause:* Poor communication between architects and engineers, leading to faulty design of the masonry wall (*ENR*, 1990a).
- Overcalculation of flows for a storm water master plan, resulting in oversizing of facilities for a large, unnamed metropolitan area in the 1990s (caused an engineering firm to incur cost overruns and tarnish its reputation). *Cause:* Failure by experts to supervise use of computer models (Hodge, 1997).
- Destruction of Mars Climate Orbiter in 1999 as a result of its going off course, causing a $125,000,000 loss. *Cause:* Failure to correctly reconcile units of measurement between vendor and owner (Heery, 2001).

Know the four non-technical factors that set the scene for negligence (NSPE, 2002)

▶ These factors were identified based on studies by the liability insurer DPIC. All four factors can be addressed by enlightened leadership and management.
- Communication
- Project team capabilities and performance
- Client selection
- Contract clauses

Use appropriate, achievable language in contracts and agreements, as shown in these examples (Hayden, 1987)

Do Not Use	Potential Replacement
At all times	Will be done once per …
Insure; ensure; assure	Reasonable effort will be made
Periodically	Not less than once per …
Supervise; inspect	Observe and report
Certify; warrant; guarantee	Statement as to our judgment based on …

Don't "let the tail wag the dog"; that is, don't allow fear of liability to prevent you from innovating and taking on new types of projects

▶ A 1999–2000 survey by the then American Consulting Engineers Council (ACEC, now known as the American Council of Engineering Companies) found that two-thirds of the responding firms reported that liability concerns "very much" or "somewhat" restricted innovation (Wallem and Salimando, 2000).

The big print giveth and the small print taketh away.

H. Jackson Brown, Jr.

▶ Counter excessive circumspection by recognizing that most liability-minimizing actions are simply sound professional and business practices that have benefits beyond addressing liability concerns. Consider these examples:

Liability-Minimizing Practice	Other Benefits to You and Your Organization
Maintaining competency	Personal satisfaction
Using standard contract forms	Save time
Writing and using guidelines	Train new personnel
Documentation	Communication within project team
Meeting deadlines	Reputation for timely service
Peer reviews	More cost-effective project for client

Read the following related lessons

▶ Lesson 31, "We Don't Make Whitewalls: Work Smarter, Not Harder"

Study one or more of the following sources cited in the lesson

▶ de Camp, L.S. 1963. *The ancient engineers.* New York: Ballantine.

▶ Delatte, N.J. 1997. "Failure case studies and ethics in engineering mechanics courses." *Journal of Professional Issues in Engineering Education and Practice* (July), pp. 111-116.

▶ Dunham, C.W., R.D. Young, and J.T. Bockrath. 1979. *Contracts, specifications and law for engineers,* 3rd Ed. New York: McGraw-Hill.

▶ *ENR.* 1988a. "Key beam under-designed" (July 7), p. 14.

▶ *ENR.* 1988b. "German bridge girder fails" (Sept. 8), p. 18.

▶ *ENR.* 1990a. "Design flaw blamed for collapse" (Jan. 18), pp. 11-12.

▶ *ENR.* 1990b. "Design led to downfall of incremental launch" (Feb. 1), pp. 13-14.

▶ Hayden, W.M., Jr. 1987. *Quality by Design Newsletter.* A/E QMA. Jacksonville, Fla. (May).

▶ Heery, J.J., Jr. 2001. "Mathematics error: when computation leads to disaster." *Engineering Times* (Oct.).

▶ Hodge, C.S. 1997. "Misunderstanding computer accuracy leads to project rework: two case studies." *Forensic Engineering: Proceedings of First Congress,* Reston, Va.: ASCE.

▶ NSPE. 2002. "New study shows structural engineers are still high-risk discipline for liability claims." *Engineering Times* (April).

▶ Walesh, S.G. 2000. *Engineering your future: the non-technical side of professional practice in engineering and other technical fields,* 2nd Ed. Reston, Va.: ASCE Press. Chapter 11: "Legal Framework."

▶ Wallem, K., and J. Salimando. 2000. "The liability noose is a bit looser." *American Consulting Engineer* (March/April), pp. 15-18.

Refer to one or more of the following supplemental sources

▶ Elovitz, K.M. 1995. "Avoiding the 10 'demandments' of contract negotiations." *Consulting – Specifying Engineer* (Sep.), pp. 15-19.

Describes clauses to not use or to be suspicious of in contract agreements and suggests alternatives. For example, "The Engineer shall perform services in a manner consistent with the highest professional services" is not acceptable. Another example: The clause "Client may terminate Engineer's service on 30 days notice" may be acceptable if the Engineer has a similar option.

▶ Weingardt, R.G. 2002. "Seeing the forest through the trees." *Structural Engineer* (July), p. 16.

Categorizes project-related peer review as either project management or technical performance. Suggests peer review generates a significant return on investment

▶ Legal columns:
- "Legal Affairs Section" in the ASCE *Journal of Professional Issues in Engineering Education and Practice.*
- "Legal Corner," column by Arthur Schwartz in the monthly NSPE publication, *Engineering Times.*
- "Legal Counsel," column by Michael J. Baker in the monthly publication, *Structural Engineer.*
- "Risky Business," column by John P. Bachner in the monthly publication, *CE News.*
- "The Law" and "Court Decisions," columns in the monthly ASCE magazine, *Civil Engineering.*

Visit one or more of these websites

▶ "ae ProNet" (http://www.aepronet.org/) is maintained by the Architects Engineers Professional Network. Included are frequently asked questions, guest essays, and continuing education and product information.

▶ "National Society of Professional Engineers" (http://www.nspe.org) provides a directory of professional liability carriers, a state-by-state summary of liability laws affecting engineering practice, and articles.

As always happens in these cases, the fault was attributed to me, the engineer, as though I had not taken all precautions to ensure the success of the work. What could I have done better?

Nonius Datus*

Nonius Datus, 152 A.D., was the Roman engineer responsible for the design and construction of a water supply tunnel through a mountain in what is now Algeria. Upon visiting the construction site he learned that the two segments of a tunnel being excavated from both ends were out of alignment and had passed each other! (de Camp, 1963)

Meetings

Meetings warrant special attention as a separate section in this book for two reasons. First, as we increase our management and leadership efforts, the number and variety of meetings we orchestrate or attend will increase. Increased involvement in meetings can be, on the balance, a positive or negative. That is, synergism happens, ideas flow, decisions occur, commitments are made, but sometimes our valuable time is wasted or at least poorly used. What we do before, during, and after meetings determines whether our increased involvement in meetings will be positive or negative, both for ourselves and others.

Second—and this may seem harsh but I believe it to be true—the majority of meetings within engineering circles are poorly planned and carelessly conducted, and they suffer from poor follow-up. Within our own circle of influence, any one of us can greatly improve meeting effectiveness and create many more positive meeting experiences for all involved.

This section presents simple techniques for enabling diverse individuals with a common purpose to function as a group, to share, debate, create, decide, and act. The suggestions are applicable to traditional, face-to-face meetings; conference calls; and multi-location, audio-video gatherings.

39

An "Unhidden" Agenda

Meetings have become the practical alternative to work.

Robert Kriegel and David Brandt

---

*T*he success of most meetings is usually determined before they start. Why? Because pre-meeting planning, especially establishing the content and tone of the agenda, determines the degree of participation and the quality of the resulting decisions.

Part of the planning should include a conscious decision about whether a meeting is needed. Perhaps some other form of communication, such as e-mail exchange, will suffice. When in doubt, don't meet.

Planning is the key to successful meetings. The meeting leader, explicitly or implicitly designated, assisted by some other likely meeting participants, should prepare the agenda. Listed below are ideas to consider as you prepare the agenda for your next meeting (Walesh, 2000). These suggestions apply to traditional face-to-face meetings, as well as to conference calls and meetings that combine conference calls for audio with website access for sharing of images.

Meetings are a lot like the hot air they produce: they'll expand or contract to fill the space available.

Robert Kriegel and David Brandt

► List all the invitees in the "To" portion of the memorandum or e-mail. Knowing who else will attend a meeting can be useful. For example, this information could provide an opportunity to informally discuss other matters with one or more individuals immediately before or after the meeting.

► Show meeting starting time and ending time on the agenda. Knowing when a meeting will end is a courtesy that enables meeting attendees to plan their day, especially the time after this meeting. A stated and firm ending time also encourages focus and brevity by participants.

► Include the entry "Additional Agenda Items" near the beginning of the agenda. This allows the leader to add topics of discussion, and it also facilitates input from attendees, some of whom may not have been involved in preparing the original agenda.

▶ Explicitly identify individuals who have reporting or other responsibilities. This helps to ensure that they will be prepared to report on their efforts or lead a discussion on the indicated topic.

▶ Attach background, support, and other materials, but don't overdo it. This extra effort by the meeting leader helps participants prepare, so that the meeting time can be used most effectively.

▶ Establish an action-oriented theme by using words and expressions such as "decide," "select," "follow up," and "select course of action." In some organizations, such as academic institutions, a meeting might be considered successful if interesting topics are discussed with no decisions made. If the meeting leader wants to avoid this, the agenda should be structured to encourage and expect action.

> *Any committee is only as good as the most knowledgeable, determined and vigorous person on it. There must be somebody who provides the flame.*
> Lady Bird Johnson

▶ Focus the energies of the participants by suggesting, for each agenda item, what participants are expected or encouraged to do. For example, indicate if the group is to develop, discuss, and select alternative solutions to a problem; or indicate that they must respond to a course of action that has been recommended by a member of the committee. Although the meeting leader must diplomatically steer the group, recognize that the group may elect to broaden its response to an agenda. However, it is unlikely to do this if trust is established between the leader and the participants.

▶ Depending on the nature of the meeting, consider adding an item entitled "Good News," under which positive happenings are briefly shared.

Being a proactive invitee can help to effect some of the preceding suggestions. For example, if you are invited to a meeting and a written agenda is not provided, ask if an agenda will follow. This may prompt preparation of an agenda and lead to a better meeting. If there will not be an agenda, insist on knowing what is to be discussed so that you can prepare, decline the invitation, or arrange to have some other, more appropriate person attend in your place.

Assume there is an important point you, as an invited participant, want to make at a meeting, but there is no directly related agenda item—either because you did not have an opportunity to get your concern on the agenda prior to the meeting, or because the chair is not receptive to adding items at the meeting. Carefully prepare your idea or information anyway. Look for opportunities at the meeting to make your point, perhaps in answer to a question posed to you or to someone else.

Suggestions for Applying Ideas

Compare these definitions of "the only two legitimate types of meetings" (Walesh, 2000) against your experience

> *Meetings are*
> *indispensable*
> *when you don't want*
> *to do anything.*
> John Kenneth Galbraith

▶ *Working meetings*. Agenda items might be problem definitions, presentation and discussion of status reports, brainstorming, conceptualizing options, comparing alternatives, and implementing solutions.

▶ *Briefings on critical non-routine topics*. Examples are personnel matters, reorganization, acquisition, or serious financial difficulties.

Decide if a meeting is really needed; that is, consider the following reasons *not* to call a meeting (Walesh, 2000)

▶ You've made up your mind what to do anyway. Convening others and pretending to seek their counsel wastes everyone's time and risks your credibility.

▶ You know what should be done, but you don't want to take responsibility. Therefore, you call a meeting, indicate what should be done, and obtain formal approval or informal acquiescence. Then you can "spread the blame" if needed.

▶ You don't know what should be done and want somebody else to make the decision. But you are responsible, it is your decision, and therefore you should decide.

▶ You're trying to use the meeting to "pull a fast one" on a colleague or your boss. In other words, he or she is out of town and is really responsible, but you call a meeting under guise of "an emergency" to make decisions on his or her project.

▶ You are on an ego trip and like the sound of your own voice.

▶ The subject is too important to merit a meeting. That is, quick, decisive action is needed and you clearly have the necessary responsibility and authority. So step up, decide, and get on with it.

▶ You've lost your case elsewhere, and you are looking for a life preserver.

Create meeting protocols or norms if your group meets regularly (based on the meeting protocol used by Patrick Engineering of Lisle, Illinois)

▶ Understand the goal of the meeting and work smart to achieve it.

▶ Prepare for the meeting. Review agenda, previous notes, and minutes. Understand the meeting's purpose and come ready to participate.

▶ Arrive promptly to avoid disrupting others.

► Bring paper and pen to take notes.

► Honor deadlines for pre- and post-meeting deliverables.

► Shut off cell phones and pager ringers or set them to vibrate. If you need to answer a call or page, do it outside the meeting room.

► Be courteous to and respectful of other attendees.

► Avoid sidebar discussions.

► Create a "parking lot" list during the meeting for issues that need resolution but are beyond the scope of the meeting agenda.

Prepare to deal with difficult behavior

► While careful meeting preparation is appreciated by participants and tends to bring out the best in them, some attendees may—unknowingly or intentionally—exhibit non-productive behavior. As a leader or facilitator of meetings, you need to recognize some of the following difficult behaviors and possibly handle them using some of the suggested solutions shown in Table 39-1 (Walesh, 2000; Ziglar, 1986).

Read the following related lessons

► Lesson 40, "Agenda Item: Good News"

► Lesson 41, "Minutes: Earning a Return on the Hours Invested in Meetings"

Study one or more of the following sources cited in this lesson

► Kriegel, R., and D. Brandt. 1996. *Sacred cows make the best burgers: paradigm-busting strategies for developing change-ready people and organizations.* New York: Warner Books.

► Walesh, S.G. 2000. *Engineering your future: the non-technical side of professional practice in engineering and other technical fields,* 2nd Ed. Reston, Va.: ASCE Press. Chapter 4, "Management of Relationships with Others."

► Ziglar, Z. 1986. *Top performance.* New York: Berkley Book, pp. 134-139.

Refer to the following supplemental source

► Hensey, M. 1991. "Keys to better meetings." *Civil Engineering* (Feb.), pp. 65-66.

Offers sensible suggestions for increasing meeting effectiveness and reducing the numbers of meetings. Some examples: speak candidly and courteously, "listen carefully enough to be able to paraphrase what was said," and expect disagreement while not taking it personally.

Table 39-1   *Difficult Behaviors in Meetings and Possible Solutions*

Difficult Behavior	*Solutions*
Person makes clearly erroneous statement.	• Offer a correction. • Indicate that the speaker is entitled to his or her opinion. • Ask others to comment on the statement.
Individual talks too much.	• Say: "Henry, I think you've made your point; let's give others an opportunity to make theirs." • Say: "Heidi, our agenda is full; we must move on." • Say: "Hans, that's interesting; but I think we are moving away from our agenda."
Person does not contribute.	• Appeal to the person's experience; say "Nancy, I understand that you have experience in this area. What is your opinion?" • Appeal to the person's position; say "John, we haven't heard from the planning department. What are your concerns?"
Individual is obstinate, continues to press for a course of action even though the group is clearly opposed, at least for the time being.	• Ask the person to put his or her ideas and arguments in writing for possible future reconsideration or for the record.
Person speaks, in a side conversation, to one or more individuals in low, possibly negative tone.	• Pause to enable others to listen. • Ask the person to repeat the comment for the benefit of the group.
Individual uses poor choice of words, erroneous terminology, or incorrect pronunciation.	• Help them by saying: "In other words, you are saying. . . ."

---

*Time spent on any item of the agenda will be
in inverse proportion to the sum involved.*

C. Northcote Parkinson

40

Agenda Item: Good News

The applause of the crowd makes the head giddy,
But the attestation of a reasonable man makes the heart glad.
Richard Steele

*D*uring an eight-year period in my career, when I managed and led a three-department unit within an organization, the three department heads and I met weekly or every other week. In keeping with my nature, we operated in a systematic fashion. Agendas were issued prior to meetings, our meetings moved right along, decisions were made, and minutes were quickly distributed and posted so that all members of the unit, and my "boss," could read them.

The first agenda item at each meeting was "Additional Agenda Items." This was done in recognition that, although meetings were carefully planned, something might "come up" at the last moment and require immediate attention.

The second agenda item for all meetings was "Good News." Typically, we quickly got to this agenda item, went around the room, and shared many specific good things that happened during the one to two weeks since the last meeting. We used this process throughout my eight years with the organization, and I cannot recall a meeting at which significant and uplifting "good news" was not reported.

Realistically, some of the good news was a "stretch." That is, one or more of us may have felt pressure, at a particular meeting, to dig deeply to find something "good" to report within our area of responsibility. However, the vast majority of good news was sincere and noteworthy. Good things were happening all around us that warranted recognition and celebration. This was how we identified the good things and the individuals and groups responsible for them.

The periodic expectation to report good news made me more aware of my surroundings. I suspect the department heads had a similar experience. In retrospect, that increased sensitivity was valuable in and of itself. It caused me, and probably the others, to listen more and talk less, to ask more questions and make fewer statements.

Continuously learning about and celebrating "good news" is an essential element in a thriving, (as opposed to surviving or dying) forward-moving organization. Once our basic physical needs are met (e.g., we earn a decent income), our work satisfaction is derived largely from challenges we face and recognition we receive. Meaningful, timely recognition requires current information regarding the good things that are happening all around us and at all levels in an organization. The "good news" agenda item is a simple, effective mechanism for obtaining this information. Try it within your circle of influence.

> *We are all imbued with the love of praise.*
> Marcus Tullius Cicero

Incidentally, you won't need a "Bad News" agenda item. That topic will take care of itself.

Suggestions for Applying Ideas

Recognize and celebrate good news through actions

▶ Document good news in widely distributed meeting minutes.

▶ Personally congratulate individuals and groups that "created" the good news by acquiring resources, completing projects, making discoveries, achieving goals, or otherwise distinguishing themselves.

▶ Publish good news on the organization's website or in its newsletter or other periodical.

Read the following related lessons

▶ Lesson 28, "Caring Isn't Coddling" (The high expectations–high support philosophy advocated will produce an abundance of good news.)

▶ Lessons 39 and 41 in the "Meetings" section of this book

▶ Lesson 45, "Our Most Important Asset"

> *He who praises everybody praises nobody.*
> Samuel Johnson

Minutes: Earning a Return on the Hours Invested in Meetings

A committee is a group that keeps minutes and loses hours.
Milton Berle

I agree with Milton Berle's basic statement: committees, in general, and their meetings, in particular, often waste time. However, I would have to disagree with even the suggestion that minutes are unnecessary. In fact, one of the reasons many meetings are much less effective than they could be is that they are not quickly followed by written documentation. Other reasons include poor planning, a weak agenda or lack thereof, excessive length, and rambling and diversions during the meeting.

In the absence of minutes, a meeting may, for all practical purposes, have never happened, that is, lots of talk but no action. Worse yet, without documentation a meeting may be remembered in as many different and conflicting ways as the number of attendees. Without the reality check provided by minutes issued shortly after a meeting, participants tend to remember what they wish had happened, not what did happen. Given the opportunity, we tend to rewrite history in our favor, editing the past so that it serves our present needs.

To avoid these kinds of complications, we should insist on minutes. These minutes should focus on decisions made and action items that resulted. Individuals who are responsible for the action items should be identified. Written minutes do not have to be long, elegant literary gems. A short, bulleted format is certainly acceptable and likely to be appreciated by recipients. My completion goal on minutes, when I prepare them or can influence their preparation, is to have them in the participants' hands or on their computer screens within three working days after the meeting.

The labor cost of preparing minutes is very small relative to the labor cost of preparing for and conducting the meeting (typically less than 5%). That small incremental investment of additional time is likely to be the catalyst for getting the desired return on investment of the large amount of time already invested in planning and conducting the meeting.

Suggestions for Applying Ideas

Record the key points of the discussion in real time, as the meeting progresses

▶ Use newsprint, transparencies, or other media readily visible to all participants.

▶ The chair, secretary, facilitator, or other specialty-designated person could provide this service.

▶ Real-time minutes are particularly useful in sensitive or controversial situations where participants want to see and review what has transpired as the discussion proceeds. Everyone has the same complete record, and everyone has the opportunity to question the recorder's interpretation.

▶ Conflicts can be resolved at the meeting as they occur instead of after the meeting when formal minutes are produced.

▶ The real-time minutes can be used to quickly prepare the traditional minutes for the meeting.

Allow participants to comment on meeting minutes—a group's success requires consensus on what was decided and why

▶ Encourage consensus by sending draft minutes to participants shortly after the meeting. Ask for review comments within an explicit time frame. Include—in the transmittal e-mail, memorandum, or letter—a note that says, "If I do not hear from you within three days, I will assume you find the draft minutes acceptable."

▶ Solicit review comments from meeting participants by immediately sending the draft minutes and then asking for corrections near the beginning of the next meeting.

Read the following related lessons

▶ Lessons 39 and 40 in the "Meetings" section of this book

▶ Lesson 34, "TEAM: Together Everyone Achieves More"

▶ Lesson 35, "Virtual Teams"

Refer to the following supplemental source

▶ Frank, M.O. 1989. *How to run a successful meeting in half the time.* New York: Simon & Schuster.

Offers many specific suggestions to improve the planning and conduct of meetings. Tools and techniques range from logistics, such as where to sit to ways to introduce ideas not on the prepared agenda.

The meeting isn't over until the paperwork is done.

Anonymous

Marketing

"Marketing" elicits intense, often negative responses from engineers. This is unfortunate given that, in the business sector, marketing is essential to survival. Its success depends on leadership by engineers or, at a minimum, enthusiastic participation by them. Marketing is also important in government, academic, and other arenas.

Lessons in this section take a positive tact. Marketing is presented as an honorable process by which some person or organization with a need connects with some person or organization that can meet that need. Win-win outcomes occur.

A Simple Professional Services Marketing Model

I don't care how much you know until I know how much you care.
Anonymous

*T*he word "marketing" often engenders negative reactions or connotations among engineers. They see images of brash, high-pressure car salespeople and are repulsed by the thought of "wasting" their professional education and experience doing "sales" work. But perhaps it is just that they have not been exposed to any particular marketing philosophy or model that inspired them to participate in the marketing function. Hopefully you will be at least receptive to the marketing model presented here. To the extent you learn to view marketing as earning trust and meeting client needs, which is the essence of the model, you may conclude that marketing is not only an ethical process but also a very satisfying and mutually beneficial one.

> *Marketing is not a department, it is your business.*
> *Harry Beckwith*

Marketing is a major expense for an engineering organization—it consumes valuable hours and dollars. Therefore, the marketing effort must be carefully planned and executed. Disciplined management and enlightened leadership are required. Professional service firms should undertake a continuous, proactive, positive marketing process, not a series of sporadic reactions "when they need work."

Stephen Covey (1990) explains that the Greek philosophy for what might now be called "win-win interpersonal and interorganizational relations" was based on ethos, pathos, and logos. In this model:

- *Ethos* "is your personal credibility, the faith people have in your integrity and competency. It's the trust that you inspire."
- *Pathos* "is the empathic side—it's the feeling. It means you are in alignment with the emotional thrust of another's communication."
- *Logos* "is the logic, the reasoning part of the presentation."

Covey emphasizes that these three elements of win-win interpersonal and interorganizational relations must occur in the indicated order. The ethos-pathos-logos sequence offers us a positive and effective marketing model, as long as the indicated sequence is followed (Walesh, 2000):

- Earn trust
- Learn needs
- Close deal (logically follow up)

The rational tendency in interpersonal relations is to start with logos, which usually leads to less-than-satisfactory results. Engineers in particular are inclined to proceed too quickly with and rely too heavily on logic.

Each of us, as engineers striving to facilitate a mutually beneficial marketing process, should first establish trust, then understand needs, and finally follow-up logically. Once trust is earned, potential clients are likely to share their needs with us in response to our questions. If there is a match, that is, if we and our firm can meet client needs, then a logical follow-up in the form of a contract or agreement is likely to occur.

> *It takes 20 years to build a reputation, and five minutes to ruin it. If you think about that, you'll do things differently.*
> Warren Buffet

If a match does not develop between our firm and the potential client, then we should provide assistance by referring the potential client to another individual or organization. Remember that our first goal is to earn trust. Being truly helpful, by making a thoughtful referral, is one way to do that.

Many specific tools and techniques are available for implementing a marketing program. An effective set of tools and techniques must be selected for each of the three steps. The book *Engineering Your Future* (Walesh, 2000) lists 56 marketing tools and techniques applicable to engineering organizations, only one of which ("ask for contract") is selling. Management consultant and writer Peter Drucker succinctly states that "the aim of marketing is to make selling superfluous." This statement underscores the idea that selling is only one small part of marketing and suggests that if marketing is done well, sales will occur naturally.

The marketing model—earn trust, learn needs, and close deal—is simple. However, its application requires patience and perseverance and, therefore, major absolute time and elapsed time. Elapsed time for a particular marketing effort will certainly be measured in months and more likely in years.

All engineers, whether in the private or public sectors, whether providing services or products, should be aware of the marketing function. Why? Because all of us ultimately help to meet client, customer, or stakeholder needs, and the essence of marketing is learning about and meeting needs. If you are employed by an engineering consulting firm, then marketing becomes a critical success factor. It should be one of your management and leadership abilities.

Suggestions for Applying Ideas

Recognize the amount of absolute and elapsed time required for each of the three steps of the marketing model

▶ The first of the three steps, ***earning trust,*** requires by far the most absolute and elapsed time. The second step, ***learning needs,*** consumes much less time. The time needed for the third step, ***closing the deal,*** is very small compared to the time required to earn the trust and learn the needs on which the sale is based. For example:

• For seven years, I communicated with the president of an engineering firm before discovering an opportunity to be of service. Largely as a result of the relationship that had been developed, I was selected on a sole source basis for an interesting, satisfying, and financially rewarding assignment.

• Admittedly, seven years is a long time to earn trust and learn needs. However, when marketing engineering and related professional services, don't expect to accomplish the personal process in weeks or even months. Select potential clients carefully and then practice patience and perseverance.

Contrast the approaches in Table 42-1 that work with approaches that do not work in the three-step earn trust–learn needs–close deal marketing model

Apply the model—earn trust–learn needs–close deal—to other situations

▶ As noted in the essay, the model is based on a win-win approach to interpersonal and interorganizational relations.

▶ Successful interpersonal and interorganizational relations and accomplishments are based on earned trust and understood needs.

▶ The next time we are confronted with a family, neighborhood, community, or other issues, ask:

• Have the principal individuals earned each other's trust? If not, work on it, because without trust the resolution of the issue will be, at best, fragile and highly legalistic.

• Do the principal individuals understand each other's needs? If not, work on it, because an acceptable resolution must try to satisfy needs.

• Can we close the deal, that is, agree to a solution or course of action? Assuming trust has been earned and needs have been learned, the answer is likely to be yes!

Read the following related lessons

▶ Lessons 43 and 44, also in the "Marketing" section of this book

Table 42-1   *Marketing Approaches That Work and Ones That Don't*

What Works	What Doesn't Work
Listening to earn trust and learn needs	Talking about what we do
Building relationships	Pursuing projects
Asking questions	Pontificating
Researching, qualifying, and ranking potential clients—a rifle approach	Viewing clients as being the same—a shotgun approach
Active involvement in targeted professional/business organizations	Passive membership in randomly selected professional/business organizations
Keeping current, both technically and otherwise	Maintaining status quo
"Face time"	Mass mailings
What you see is what you get	Bait and switch
Illustrating benefits	Pushing features
Multiple level contacts with client	Single level contact
Suggesting program and project approaches	Reacting to requests for proposals
Tailoring to client	Boilerplating from files
Defining and meeting requirements	Talking "quality" and spewing slogans
Delivering locally while drawing globally	Attempting to do it all locally
Admitting errors and fixing them	Blaming others
Caring for existing clients—performing on their projects	Neglecting existing clients—chasing new ones
Perseverance	Instant success
Saying "thank you"	Being presumptuous

Study one or more of the following sources cited in this lesson

▶ Covey, S.R. 1990. *The 7 habits of highly effective people: restoring the character ethic.* New York: Simon & Schuster.

▶ Walesh, S.G. 2000. *Engineering your future: the non-technical side of professional practice in engineering and technical fields,* 2nd Ed. Reston, Va.: ASCE Press. Chapter 14, "Marketing Technical Services," develops and describes the earn trust–learn needs–close deal marketing model.

Refer to the following supplemental source

▶ Lantos, P.P. 1998. "Marketing 101: how I got my ten largest assignments." *Journal of Management Consulting* (Nov.), pp. 38-40.

Outlines a variety of circumstances that led to ten contracts for a management consultant. Trust is explicitly mentioned in six and implicit in the others. Also evident is the need for a proactive and varied approach.

Subscribe to one or more of these e-newsletters

▶ "TelE-Sales Hot Tips of the Week," a free e-newsletter provided by Art Sobczak, typically includes practical sales topics that complement the marketing model presented in this lesson. To subscribe, go to http://www.businessby phone.com/backissues.htm.

▶ The "LawMarketing Newsletter," a free e-newsletter offered by Larry Bodine, a marketing and web consultant. Provides marketing ideas, typically in the form of short pragmatic articles. Although the e-newsletter is focused on attorneys, it also is potentially useful to engineers, which should not be surprising given the wide applicability of the basic principles underlying effective marketing. Examples of topics addressed in recent editions of this e-newsletter are marketing in slow times, cross-selling, and creativity. To subscribe, go to http://www.lawmarketing.com/.

What you do with your billable time determines your current income,
but what you do with your non-billable time determines your future.

David Maister

43

Speed as a Competitive Edge

*In an increasingly networked economy,
it's not the big that beat the small, but the fast that beat the slow.*
John Chambers

Several years ago, after careful research, my wife and I found just the right boat, immediately made an offer, and needed to quickly arrange financing. I called our bank, explained the situation, and asked for a financing commitment. I received the traditional response: "the committee will look into it." But much to my surprise, the bank officer called back in 20 minutes, indicated that they would provide financing, and said processing the paperwork would be a formality. This was a pleasant "quickness-of-service" experience.

At about that same time, I began using one of those "quick oil change" places. The attraction was the promise of 10-minute service. This commitment contrasted sharply with my usual "service station," which typically said something like, "If you have your car into our place by 9:00 a.m., we should be able to have it for you by the end of the day."

At the "quick place," I could see the technicians work on the car—they exhibited enthusiasm and looked like they knew what they were doing. Based on the computerized list of services I received, they also did a lot of things that extended way beyond changing the oil and filter, such as adjusting tire pressures and all the fluid levels. This is reassuring. Finally, I don't pay any more, at least not much more, then I would at the "service station." Another example of speedier service.

I once dropped dress shirts off at the cleaners at about 9:30 a.m. The conversation went something like this: "When I asked when the shirts would be ready, the clerk answered, "After 3:00 p.m." "But on what day?" I asked. "Today!" she replied. I asked why they were trying to turn the work around so quickly. She said, "We've adopted a goal of in and out on the same day to provide even better service than our competitors." Another example of faster service.

Within a week, I took a sport coat to another dry cleaner for cleaning and for minor tailoring. The clerk informed me that the coat would be ready in one week. Having been recently sensitized to quickness of service, I recalled that everything has always taken a week at this place. "Why do you need a week to alter and dry clean a sport coat—surely this requires only a few hours?" I asked. The clerk replied, "We have a week's backlog."

Her boss, on overhearing our conversation, asked with concern, "Is there a problem, sir?" "No," I replied, "I was just trying to understand why things take so long, especially compared to the other cleaner that I normally use." The boss assured me that everything was normal. "We've been backed up one week for at least seven years."

Putting on my pro bono consulting hat, I suggested that, after a crash catch-up effort, they could offer one- or two-day cleaning and alteration services to set themselves apart from many other similar businesses. This suggestion was greeted with two strange looks—they just didn't get it.

And what about the consulting engineering business and the operation of governmental units? Here are some examples:

- Why do we use three weeks to prepare a proposal in response to a request for proposal (RFP) when 90% of the effort is expended in the last one to two days? We could do this in less elapsed and absolute time and at less cost.

- Why do meeting minutes always seem to come out a week or more after a meeting, if at all, especially when immediate preparation and distribution of minutes would motivate and enable meeting participants to move ahead with assigned or volunteered action items?

- Why do highly disruptive urban public works construction projects drag on for months and months, with significant periods of no activity, resulting in unneeded and unnecessary citizen frustration?

Given that banks, auto lubrication shops, cleaners, and other businesses earn a competitive edge with quickness, maybe quickness could be the distinguishing benefit offered by some consulting businesses and government entities. Fast delivery might be especially useful for firms that provide basic, commodity-type services that can easily be obtained elsewhere. Quickness could differentiate these commodity firms from competitors. Although government entities don't need quickness for a competitive advantage, speedier delivery of routine services might improve relations with citizens while reducing taxes and fees.

> *Unless we hasten,*
> *we shall be left behind.*
> Lucius Annaeus Seneca

Ingredients needed to speed up processes include self and organizational discipline, identification and use of best practices, and effective use of computers and other electronic production and communication tools. Think about the advantages of quickness during the 10 minutes needed for your next oil change. Some of the speedy service tactics they use may be transferable to your work environment.

Suggestions for Applying Ideas

Apply benchmarking as a means of identifying possible ways to increase your organization's quickness

▶ Benchmarking is "the continuous process of measuring products, services, and practices against the toughest competitors" (Camp, 1989). More simply stated, benchmarking means learning from the practices of others, whether you or they are in the public, private, or volunteer sectors.

▶ Identify a process in your organization that needs to be done faster.

▶ Look for organizations that excel in doing your or similar processes quickly. Don't necessarily confine your search to competitors or even similar organizations. For example, what could your engineering company, manufacturing firm, or government entity learn about speed from the way Dell Computers quickly handles laptop repairs, the way Oil n' Go provides oil changes and related services, and the way FedEx delivers packages overnight?

> *In skating over thin ice our safety is in our speed.*
>
> Ralph Waldo Emerson

▶ Learn as much as you can about the organizations you identify and, more specifically, their quick processes. Possible sources of information include articles published about the processes, the organizations' websites, asking the organizations for help, and retaining a consultant.

▶ Integrate some of what you learn into your processes and monitor the effects on your operations.

Read the following related lessons

▶ Lessons 43 and 44, also in the "Marketing" section of this book

Study the following source cited in this lesson

▶ Camp, R.C. 1989. *Benchmarking: the search for industry best practices that lead to superior performance.* Milwaukee, Wis.: ASQ Quality Press.

> *Better three hours too soon,*
> *Than one minute too late.*
>
> William Shakespeare

44

The Chimney Sweep and the Sewer Cleaner: The Importance of Style

Every production of genius must be the production of enthusiasm.
Benjamin Disraeli

*O*nly the owner of an "older home" can fully appreciate having everything go wrong at once. This time, "everything" included sewer and chimney problems. Tree roots had apparently once again plugged the lateral sewer. Besides a good cleaning, the fireplace chimney needed a screened cap to keep out the squirrels.

(The last bushy-tailed visitor dropped onto the hearth shortly after we started a fire, sped to the dining room, and frantically jumped onto the windowsill. We opened the front door and, after taking one lap around the dining room table, the squirrel ran outside to freedom. That was enough!)

The chimney sweep and the sewer cleaner were scheduled for the same day. The chimney sweep arrived, strode directly to the front door, and rang the bell. Somewhat to my wife's surprise, he was formally attired: top hat, white shirt, bow tie, and black coat with tails.

He politely introduced himself and responded to my wife's curiosity by explaining the history of chimney sweeps and, in particular, their garb. Chimney sweeps were of the poorest class in Europe. They depended on castoffs for their clothing and often acquired the discarded formal attire of the undertakers. After excusing himself, the chimney sweep began his initial inspection of our chimney.

As the chimney sweep walked away from the front door, the sewer cleaner drove his truck into the driveway. He trudged around the house to the back door. My wife answered the doorbell and noticed that the sewer cleaner's clothing, in contrast to the chimney sweep's, was strictly functional—green work clothes and heavy boots. The sewer cleaner didn't bother to introduce himself. Instead, he mumbled something about the "weirdo" at the front door, and then went down into the basement to begin his work.

The chimney sweep completed his initial assessment and returned several days later with a taller ladder and special cleaning equipment. He was up and down the ladder, in and out of the house, and then back up the ladder as he went about cleaning the chimney and installing the cap.

Because our home was on a busy street, the chimney sweep attracted considerable attention, and many passing motorists sounded their horns. During one of his trips into the house, the chimney sweep explained that the proper response to greetings from the passersby was a tip of the hat and bow from his position at the top of the ladder. However, he was frustrated because the traffic was so heavy and the beeping so persistent that he simply could not take the time to give the traditional tip of the hat and bow. Therefore, he compromised and simply waved. After completing his work, the chimney sweep presented his bill and politely said good-bye.

Chimney work is probably no more or no less important than sewer work. In terms of desirability and prestige, both trades would probably rank low. And yet, there was something special about the way our chimney was cleaned compared to the way our sewer was unplugged. I suspect that the cheery, enthusiastic chimney sweep felt better about cleaning the chimney than the glum sewer cleaner felt about unplugging the sewer. While both the chimney sweep and the sewer cleaner provided valuable services, the chimney sweep did it in such a way so as to bring a bit of cheer to us and to the many passersby.

This story illustrates an important point. Although the work we do is important, the manner in which we do it significantly affects the way our efforts are received and appreciated by others. Think of your favorite restaurant, hardware store, or hair stylist. While the quality and price of products or services help to define "favorite," I suspect that the attitude of employees, combined with the physical setting, enters into the equation.

> *Enthusiasts soon*
> *understand each other.*
> Washington Irving

The same style principle applies to engineers, whether we are in businesses that provide services to clients or customers or in a government entity assisting citizens. More specifically, style enhances marketing like frosting on a cake or the bow on a package. As we seek new clients and customers, and strive to improve service to existing ones, let's explicitly include style in the effort. Exude enthusiasm, be polite, listen carefully, speak clearly, explain thoughtfully, assist positively, dress appropriately, walk tall, and smile!

Suggestions for Applying Ideas

Assess the style of your organization, or the portion for which you are responsible

▶ Try to imagine you are a client, customer, or other stakeholder. Forget what you know about your workplace and the people who work there.

- "Call yourself up," suggests Robert Townsend (1970), the former CEO of the Avis rental car company. Call your office. What is your perception? If it's negative, how could it be fixed?
- Approach and enter your building. Do the external and internal physical environment speak of what you do and how you do it? Does the setting convey the importance of your organization's work and the pride you take in it? Or is it generic and bland?

▶ Better yet, eliminate the need to imagine you are an outsider. Ask a friend who knows little about your organization to call or visit.

▶ Consider this anecdote: While serving as engineering dean, a prospective student and her mother met with me as part of a visit to our campus. They had already visited at least one other campus. The mother immediately said she really liked the school compared to the one she visited the day before. Obviously, I asked why. Her answer: The entrance to the engineering building was neat and clean. Of course, faculty, curricula, and laboratories are important, but so is the style of the place.

> *More often than not, things and people are as they appear.*
> *Malcolm Forbes*

Read the following related lessons

▶ Lessons 42 and 43, also in the "Marketing" section of this book

Study the following source cited in this lesson

▶ Townsend, R. 1970. *Up the organization: how to stop corporations from stifling people and strangling profits.* New York: Alfred A. Knopf.

> *Promotion awaits the employee who radiates cheerfulness, not the employee who spreads gloom and dissatisfaction.*
> *B.C. Forbes*

Building Mutually Beneficial Employee–Employer Partnerships

Question for employers: What goes down the elevator and out the door every day? Answer: Your most important "asset"—your personnel.

Question for employees: As you go down the elevator and out the door, what determines whether or not you look forward to returning tomorrow? Answer: Probably a mixture of tangible and intangible factors that define the success and significance of your work.

The old employee–employer contract, under which the former toils faithfully and the latter guarantees a job, is largely gone. In its place is a more sophisticated, challenging, and improved model under which both parties seek deep and broad benefits and each is free to discontinue the relationship if its productivity declines.

Employers need to examine their recruitment and retention process to recognize today's realities. Similarly, prospective employees should carefully conduct their job searches. Lessons in this section offer advice to both employees and employers.

45

Our Most Important Asset

> *What counts in any system is the intelligence, self-control,*
> *conscience, and energy of the individual.*
>
> Cyrus Eaton

Near the end of our telephone conversation, my client noted that six people in his office—a large consulting firm—were just "let go." He went on to say that some were very capable, but backlog was down and time utilization was slipping. Reducing personnel would fix both problems, at least in the short run. The day before, in a conversation with another representative of the same firm, but in another office, I heard how hard it is to find good people.

As an outsider looking in, I have seen this scenario again and again. Let people go at one location, for a short-term bottom-line fix, while trying to deal with severe personnel needs at other locations. I understand the realities of personnel being at the wrong location or in the wrong discipline to meet current needs. But isn't there a better way to practice stewardship with our most valuable resource—our engineers and our technical and other personnel?

With competent, creative, and conscientious people, our organizations can do almost anything. Without such people, regardless of what other resources we may have, we will struggle simply to survive.

In my view, enlightened education and training (E&T) is part of the answer to effective utilization of personnel resources for individual and organizational benefit. Consider these suggestions for building an E&T program:

- Audit your E&T efforts. Retain a consultant to help you take a fresh look at how you spend resources. Are you cost-effectively investing resources (time and dollars) so that they yield an attractive return on investment (ROI) by meeting your organization's technical, marketing, and other needs? Or is E&T just another expense, and poorly managed at that?
- Partner with your personnel on E&T. Both you and they should invest time and dollars for maximum ROI.

- Experiment with various teaching and learning mechanisms, including rapidly emerging distance-learning technologies, such as web-based E&T. Younger personnel, in particular, are likely to embrace these.
- Leverage your E&T investments. For example, require some form of reporting, sharing, or action from each person who participates in any learning activity.
- Implement coaching (easy) or mentoring (difficult) programs as a proactive way for senior personnel to share knowledge with junior personnel.

Besides E&T, other enlightened means are available to retain and cultivate high-quality personnel. One is to recognize individual and group technical and non-technical achievements. We work for different reasons, but most of us value timely and sincere private and public recognition. Other ways to care for our people assets include profit sharing, purchasing stock, challenging assignments, pleasant physical surroundings, modern equipment, dual advancement ladders, flextime, and flexible benefits. Finally, use today's exploding technology to enable underutilized personnel, regardless of their physical location, to work on project teams.

> *Making sure that everyone in the organization knows exactly what his job is and what its purpose is and how it fits in— and how you know you are doing well— is an arduous and never ending process, but it's the single most important element in managing people.*
>
> John Mole

Laying off high-quality people because they don't have productive work to do reflects negatively on an organization. Leaders and managers should determine future service needs; use E&T and other means to recruit, retain, and strengthen the best and brightest personnel to meet those needs; and employ technology to enable personnel to work together regardless of their location. It's a matter of good stewardship and will also help the bottom line. Our most important assets go down the elevator or out the door every day. We must make it worthwhile for them to keep coming back.

Suggestions for Applying Ideas

Influence your organization's culture by empowering and developing its personnel, particularly if you are in a position to do so

▶ Contemplate John Mole's experience and observations as described in his book, *Management Mole: Lessons from Office Life* (1998).

- John Mole, an educated and experienced manager, quit his management position and went underground for two years as a "mole."
- During that time, he had temporary jobs in 11 organizations, jobs he obtained without revealing his education and business background.
- His disturbing observations, as presented in the book, are as follows:
  —The majority of junior-level staff he encountered wanted to learn, work, contribute, and succeed.

—Unfortunately, much of that potential was wasted because of poor management and leadership. Orientation, education, and training were virtually nonexistent. As a result, the "blind lead the blind."

▶ Is your organization tapping and enabling its people resources? Or are you letting them flounder?

Retain high-quality personnel by implementing some of the suggestions listed below

▶ Select and try those that are most likely to fit your organizational culture. Recognize that tactics that retain top personnel within your organization also tend to attract top personnel to your organization. (This list contains suggestions from most of the publications cited in the sources section of this lesson.)

▶ Create a *young leaders group* that meets periodically with various senior personnel to discuss and learn about professional practice, business, and your organization.

▶ Develop a *seminar/workshop series*, to complement the preceding, in which participants would progress through a series of management and leadership modules. The series could be led mostly by your personnel with occasional participation by outside consultants.

▶ Conduct *focus groups/forums* to remain current with respect to personnel issues and concerns.

▶ Establish an anonymous, online *question/suggestion system.*

▶ Encourage more senior personnel to *mentor and/or coach* junior personnel as a means of providing more significance for seniors and more attachment for juniors.

▶ Help each person design a multi-year *personal development plan* in which your organization participates, to the extent feasible, by providing supportive assignments and education and training.

▶ Institute *teleworking* and/or *flextime.*

▶ Provide *child care, eldercare,* and other *special services.*

▶ Grant *retention bonuses.*

▶ *Recognize* individual and team professional achievements.

▶ Institute optional or required *360-degree feedback.*

▶ *Link education and training* to your organization's operations. For example, conduct a reengineering workshop to explain fundamentals and then apply some to a real organizational process. Application-oriented E&T tends to attract and hold bright personnel.

▶ Provide technical and managerial advancement *routes.*

▶ Recognize the *profile of Generation Xers,* that is, employees ranging in age from about 25 to 40. They tend to want to participate, expect quick results, want frequent feedback, value non-work life, want to build a resume, expect fun, and are skeptical of seniors.

▶ Recognize that employees may *pass through* these four *phases,* starting when they enter your organization and continuing as they progress: introduction, growth, maturity, and decline. Each requires special measures.

▶ Inform personnel about *open positions* throughout the organization.

▶ *Rotate* personnel through various roles and functions, such as planning through construction.

▶ *Automate* routine functions to reduce tedium.

▶ Provide *relocation bonuses.*

▶ Recognize that even sophisticated work can become *boring*.

▶ Support active (not passive) involvement in professional *societies* and/or business *organizations.*

> *Hire the best.*
> *Pay them fairly.*
> *Communicate*
> *frequently. Provide*
> *challenges and rewards.*
> *Believe in them.*
> *Get out of their way*
> *and they'll knock*
> *your socks off.*
> Mary Ann Allison

▶ Change the behavior of or remove *negative/incompetent personnel.*

▶ Explain the *application/value* of each assignment.

▶ Rectify *unsatisfactory relationships* between employees and supervisors/managers.

▶ Reduce *bureaucracy.*

▶ Offer ever-increasing *responsibility.*

▶ Review your *recruitment process* to increase the likelihood of finding compatible personnel. Seek clear communication of mutual expectations and determine compatibility of each candidate with the organizational culture.

▶ Pay for *performance.*

▶ Recognize that while some employees may remain with your organization, you may be gradually *losing them psychologically;* their interest and energy may be declining.

Use "people assets" to help your organization rise to greatness

▶ Management researcher Jim Collins, in his book *Good to Great* (2001), observes that good companies rise to greatness because of their people. Some other observations:

- "People are not your most important asset. The right people are." This statement supports and gives added emphasis to the theme of this lesson.

- Individuals who led companies that ascended from good to great almost all shared two traits: personal humility and professional will. "Their ambition is first and foremost for the institution, not themselves." So much for charisma?

- The most effective leaders "look out the window to apportion credit. . . . At the same time, they look in the mirror to apportion responsibility."

> *The purpose of bureaucracy is to compensate for incompetence and lack of discipline— a problem that largely goes away if you have the right people in the first place.*
>
> *Jim Collins*

- Somewhat surprisingly, the path from good to great did not typically begin with a vision and strategy. It did not start with deciding "where to drive the bus." Instead, successful leaders "first got the right people on the bus and the wrong people off the bus" and then figured where to drive it. Stated differently, begin with "who," then address the "what."
- "Put your best people in your biggest opportunities, not your biggest problems." This advice counters the tendency to invest the knowledge, skills, and attitudes of our star personnel in fixing big messes created by others.
- Don't waste time and energy trying to motivate people. Instead, invest time and energy in finding and hiring motivated people and work hard to support them, to not demotivate them.

Read the following related lessons

▶ Lesson 34, "TEAM: Together Everyone Achieves More"

▶ Lessons 46 and 47, also in Part 7 of this book

Study one or more of the following sources cited in this lesson

▶ Allen, C. 2001. "Using the power of diversity to retain staff: developing tools to ensure success." *Leadership and Management in Engineering* (Winter), pp. 22-25.

▶ Avila, E.A. 2001. "Competitive forces that drive civil engineer recruitment and retention." *Leadership and Management in Engineering* (July), pp. 17-22.

▶ Collins, J. 2001. *Good to great: why some companies make the leap and others don't.* New York: HarperCollins.

▶ Forrest, D.J. 1999. "Employer attitude: the foundation of employee retention." *Keep Employees, Inc.* (Dec.) http://www.keepemployees.com/WhitePapers/attitude.pdf.

▶ Galloway, P.D. 2001. "Innovative benefits in a small consulting firm." *Leadership and Management in Engineering* (Winter), pp. 45-47.

▶ Glagola, C.R.F., and C. Nichols. 2001. "Recruitment and retention of civil engineers in Department of Transportation." *Leadership and Management in Engineering* (Winter), pp. 30-36.

▶ Hensey, M. 2001. "Innovations and best practices: leadership development and retention." *Leadership and Management in Engineering* (Winter), pp. 37-41.

▶ Hessen, C.N., and B.J. Lewis. 2001. "Steps you can take to hire, keep and inspire Generation Xers." *Leadership and Management in Engineering* (Winter), pp. 42-44.

▶ Mole, J. 1998. *Management mole: lessons from office life.* London: Bantam Press. [For a review, see Bredin, J. 1988. "Confessions of a management mole." *Industry Week* (Sept. 19), p. 32.]

▶ Morrell, K., and M. Simonetto. 1999. "Managing retention at DeLoitte Consulting." *Journal of Management Consulting* (May), pp. 55-60.

▶ Pitzrick, D.A. 2001. "One company's approach to recruitment and retention." *Leadership and Management in Engineering* (Winter), pp. 48-50.

▶ Rosenbluth, H.F., and D.M. Peters. 2002. *The customer comes second.* New York: HarperCollins. (The authors note that "An ounce of retention is worth a pound of recruitment.")

▶ Smither, L. 2003. "Managing employee life cycles to improve labor retention." *Leadership and Management in Engineering* (Jan.), pp. 19-23.

Refer to the following supplemental source

▶ Walesh, S.G. 2000. *Engineering your future: the non-technical side of professional practice in engineering and other technical fields,* 2nd Ed. Reston, Va.: ASCE Press.

Chapter 4, "Management of Relationships with Others," includes a subsection entitled "Appreciating and Working with Support Personnel." Topics presented are essentiality of support personnel, challenges unique to working with support personnel, and tips for working effectively with support personnel.

Why spend all that money and time on the selection of people when the people you've got are breaking down from underuse?

Robert Townsend

46

Interviewing So Who You Get Is Who You Want

Remember: A's hire A's and B's hire C's.

Donald Rumsfeld

A consulting firm client asked me to assist with interviewing a final candidate for a project manager position. I studied the candidate's resume and interviewed him over lunch. I asked many questions, such as positions he had held, the types of projects he worked on, and software he had used. To me, he looked great on paper and in person! He was hired. To my dismay, and my client's, the new project manager quickly and clearly demonstrated his inability to write letters, memoranda, and reports that could be understood by anyone other than perhaps himself.

The employer generally gets the employees he deserves.

J. Paul Getty

Besides embarrassing me, this interviewing experience crystallized for me the importance of a carefully designed interview process. Finding individuals whose producer, manager, and leader profile match the culture and needs of your organization is a challenge. Accordingly, a thorough and systematic process is needed. The process, which should begin with careful study of the resumes and other submittals and reference checks, might include all or most of the following five steps:

1. Define, in writing, knowledge, skills, and attitudes needed for the position. Call these the criteria. This step, assuming it is a team effort, may reveal unexpected, widely divergent views on "what we are looking for." If so, get consensus before proceeding. The object of the interview is to determine the degree to which a candidate satisfies the criteria.

2. Arrange, to the extent feasible, for the candidate to explicitly demonstrate compatibility with the criteria during the interview visit. Examples of criteria that can be demonstrated during an interview are writing, speaking, and problem assessment and solution.

3. Consider reviewing examples of work products as a way of assessing a candidate's ability to meet some knowledge, skills, and attitudes criteria. Assume a criterion is creating highly communicative graphics. Ask the candidate to bring some examples. If the ability to design aesthetically pleasing public works facilities is desired, request the candidate to bring photographs of some completed projects.

4. Use what is sometimes called behavioral interviewing for those criteria not amenable to explicit demonstration—or "show and tell"—during the interview. With this approach, the candidate is asked to relate actual personal, and hopefully revealing, experiences that illustrate desired attributes. Assuming the responses honestly portray his or her behavior, this technique is based on the premise that recent behavior is the best predictor of near future behavior.

5. Recognize that positive interpersonal "chemistry" is an important, but often unstated, criterion in personnel selection. Accordingly, a variety of independently derived views of the candidate's potential team members and colleagues are desired. Schedule private, one-on-one discussions between the candidate and representative members of your organization's staff. Encourage both parties to be open and direct. Ask each involved staff member to brief the leader of the interview effort in person and/or in writing, immediately after the discussion.

It's much less expensive to recruit from the top of the barrel by paying top wages.
Robert Townsend

A thorough interviewing process will put you in an excellent position to answer a very important question: Which candidate is most likely to enter into a mutually beneficial relationship with our organization? You and others will now know. You have done your homework.

Suggestions for Applying Ideas

To the extent feasible, have candidates demonstrate compatibility with the position criteria during the interview visit (step 2 of the process described in the lesson)

▶ If writing ability is one criterion for a government position, obtain a sample of the candidate's writing during the interview, preferably near the end of the on-site visit. Invite the candidate to write about ways in which personal experience and aspirations are in sync with the available position and the government entity's mission. Provide a quiet spot with paper or computer, and allot about one-half hour. A good to excellent writer will shine on this essay "test."

▶ If speaking skill is crucial, request that the candidate make a presentation or conduct a workshop on a topic related to the available position. Include this requirement in the interview invitation, and indicate the allotted time. I once worked for an engineering college that required candidates to deliver a typical student-oriented lecture, with students and faculty in attendance, as part of the interview visit. Your candidate could describe a completed design or construction project; demonstrate software he or she developed; or teach participants, in a workshop mode, how to analyze geotechnical data.

> *There is something that is much more scarce, something rarer than ability. It is the ability to recognize ability.*
> *Robert Half*

► If proposal preparation ability is critical for a consulting firm position, give the candidate an actual or hypothetical request for proposal and about one hour of quiet time. Ask him or her to list questions that should be answered by the firm's personnel or the client for two situations: before a "go, no-go" decision is made, and, if a "go," during the proposal preparation process.

Use behavioral interviewing (step 4 of the process described in the lesson)

► For example, assume that persistence is a desired quality. Ask the interviewee to cite a personal experience that demonstrates persistence.

► You can easily imagine using this behavioral approach to check out other potentially desirable qualities, such as creativity, leadership, and teamwork.

► The retrospective, reality-based, behavioral approach contrasts sharply with the prospective, hypothetical approach. An example of the latter is, "What would you do to encourage out-of-the-box thinking on your team?" An example of the behavioral approach is, "Give me an example of how you encouraged out-of-the-box thinking on your team?" Slightly different wording, markedly different question.

► Concrete, historic examples reveal much more than hypothetical projections. Knowing what someone did do is much more valuable than what someone says they would do.

> *Be mindful of candidates' attitudes; they are as important as skills and experience, and harder to change.*
> *Mel Hensey*

Discern the leadership potential of candidates by looking for their willingness to do the following (Reiland, no date)

► Follow—Reveals their attitude.

► Sacrifice—Reveals their perspective on life.

► Learn—Reveals the condition of their ego.

► Serve—Reveals their heart.

► Be honest—Reveals their maturity.

Institute programs in the organization that are known to attract high-quality personnel, in addition to meeting other purposes

► A progressive education and training program will tend to attract "top" personnel. High-quality professionals want to maintain their competence and, therefore, tend to join and remain in organizations that value and support education and training (Meister, 1994; O'Connell, 1996).

▶ Participating in cooperative education is another way to attract "top" new graduates. "Co-op" integrates classroom studies with planned, paid, and supervised paraprofessional work experience in the public and private sectors. Based on my educator and practitioner experiences, co-op provides access to top personnel because, when given a choice, the better students tend to elect the co-op option. Employers get a very close look at their co-op students and can evaluate them for possible full-time employment. Similarly, co-op students can assess the employer. As a result, better matches occur.

Avoid these hiring pitfalls (Tobias, 1987)

▶ "Hiring expediently under the pressure of time. That's the 'buy now, pay later' approach."

▶ "Resorting to hiring the 'best of the batch' out of desperation."

▶ Compromising standards by arguing, "Don't some of us have to be Indians and not chiefs?"

▶ Hiring castoffs of superb organizations or the stars of mediocre organizations.

▶ Not liking what is seen but "hiring with the hope the person will change" once they come onboard. Is the unmotivated person likely to become motivated? Is the underachiever likely to achieve? Is the narrowly focused individual likely to expand his or her horizon? I doubt it.

▶ Failing to determine if the candidate has the necessary knowledge and skills.

▶ Weighing knowledge and skills too much relative to attitude.

▶ Hiring without carefully checking credentials and prior performance.

▶ Relying too much on answers to prospective, hypothetical questions and too little on responses to the retrospective, behavioral questions.

▶ "Ignoring your personal feelings or gut reactions."

▶ Insisting on cloning yourself.

Read the following related lessons

▶ Lessons 45 and 47, also in Part 7 of this book.

Study one or more of the following sources cited in this lesson

▶ Freese and Nichols. No date. "Freese and Nichols University: Where Education Is a Tradition." (Brochure published by Freese and Nichols, a consulting firm that uses its corporate university as a recruitment tool.)

▶ Meister, J.C. 1994. *Corporate quality universities*. New York: Irwin Professional Publishing. (The author states that "the ability to offer all levels of employee opportunities to continuously learn new skills . . . will be a crucial factor in attracting and maintaining a first-class work force.")

▶ O'Connell, M. 1996. "Training as a potential profit center." *Journal of Management in Engineering* (Sep./Oct.), pp. 25-27. (Benefits of the consulting firms SEH's progressive education and training program included enhanced recruiting and low turnover.)

▶ Reiland, D. October 6, 2003. "The Pastor's Coach," monthly e-newsletter available at www.INJOY.com. The list is used by permission of Dr. Dan Reiland.

▶ Tobias, L.L. 1987. "Hiring for excellence." *Industry Week* (April 20), p. 71.

Refer to the following supplemental source

▶ Newton, M. 2001. "Interviewing tips." *Leadership and Management in Engineering* (July), pp. 5-6.
Offers tips to help an interviewer be more effective when meeting with a prospective employee. Examples are helping the interviewee prepare, demonstrating staff camaraderie and enthusiasm, using a variety of questioning styles, and having candidates do most of the talking.

If you want a track team to win the high jump,
you find one person who can jump seven feet,
not seven people who can jump one foot.
Paul Dickson

Eagles and Turkeys

Walk with wise men and you'll be wise.
But keep company with fools and you'll suffer for it.
Prov. 13:20, The Bible, *An American Translation*

*I*f we want to fly like eagles, we cannot get our wings from turkeys. The aspiring manager and leader in us must have regular contact with individuals who are accomplished in managing and leading—the eagles. However, we can learn what not to do from the latter—the turkeys.

Think of members of the engineering community, or any community, as being represented by a normal distribution. Eagles are found at the extreme right end. The vast majority of us occupy most of the area under the curve. We are a decent, hardworking bunch. The turkeys—that is, the whiners, cynics, grumblers, complainers, talkers, incompetents, and malcontents—occupy the extreme left end of the distribution.

> *Depend on no man, on no friend but he who can depend on himself. He only who acts conscientiously toward himself, will act so toward others.*
> Johann Kaspar Lavater

Bosses, co-workers, partners, clients, and customers influence our attitude toward the engineering profession, affect the knowledge and skills we acquire, and place us in various networks. Unless you are an unusually independent, self-disciplined person, your people environment will shape you. Therefore, strive to align at least some of that people environment with your desired roles and goals; that is, try to spend some time with the eagles over at the extreme right end of the distribution.

I do not apologize for advising you to seek contact with accomplished individuals. In spite of all of this world's problems, its population includes a small minority of conscientious, competent, and communicative managers and leaders—visionaries who act with honesty and integrity, set and achieve goals, exhibit courage, accommodate ambiguity and chaos, and are innovative and creative. Celebrate their presence, associate with them, and learn from them.

> *Thus you will know them by their fruits.*
> Matt. 7:20, The Bible, RSV

Sure, we can learn about managing and leading through formal education, by reading the literature, and as a result of our mistakes. However, our learning will be accelerated by frequent working contact with individuals who possess the knowledge, skills, and attitudes we desire.

Just as real eagles (i.e., the majestic birds) are out there but hard to find, so is the case with those accomplished in managing and leading. How do we find these people? Three suggestions:

- Look within all types of organizations and groups such as businesses, government entities, volunteer groups, service clubs, business and professional societies, and neighborhood associations. Do not limit your search to engineering groups.
- Scan these organizations and groups vertically. Appreciate that the most accomplished managers may be closer to the top of the organizations and groups and may hold management titles.
- Recognize that some of the more accomplished leaders—those who influence primarily by their presence rather than positions—may not be immediately identified. They tend to eschew titles, trappings, and outward signs. Instead, they are widely known by their positive, supportive influences on others. Simply ask the "others" who they look up to. If leadership is present, you will immediately find the source.

> *Keep away from people who try to belittle your ambitions. Small people always do that, but the really great make you feel that you, too, can become great.*
> Mark Twain

Having found one or more of those individuals who exhibit exemplary managing and leading ability, seek ways to have meaningful contact with them. For example, depending on the situation:

- Volunteer to serve on their task force.
- Request a transfer to their department.
- Join their business association or professional society.
- Seek their advice.

Once you've found and connected with an individual who excels in managing and leading, your real work begins. Earn his or her trust by working diligently and smart while demonstrating honesty, integrity, and competence. Then you will be amazed at what you, your new teacher, and others accomplish.

Suggestions for Applying Ideas

Carefully choose your co-workers and your boss, especially in the early, more formative part of your career

> *We've never succeeded in making a good deal with a bad person.*
> Warren Buffet

▶ Consider employment factors such as the following: geographic location, compensation, likely projects and functions, available computer and other equipment, and style/condition of office and/or work area.

▶ However, if you have high managing and leading aspirations, carefully choose your co-workers and your bosses; they are the most important factors early in your career.

As you seek people with whom you can develop mutual trust, be alert to these six character flaws (McCormack, 2000)

> *Avoid the reeking herd,*
> *shun the polluted flock*
> *live like the stoic bird*
> *the eagle on the rock.*
>
> *Elinor Wylie*

- "People who never do what they say they will do.
- People who push their work onto you.
- People who are late and don't apologize.
- People who tell you 'I'm too busy.'
- People who reject your ideas.
- People who won't let you off the hook."

▶ In other words, "Place high value on trust, but don't be too trusting too soon."

Read the following related lessons

▶ Lesson 1, "Leading, Managing, and Producing"

▶ Lesson 8, "Thank Our 50 Stars"

▶ Lessons 45 and 46, also in Part 7 of this book

Study the following source cited in the text

▶ McCormack, M.H. 2000. *Staying street smart in the Internet age: What hasn't changed about the way we do business.* New York: Viking Press.

Choose your friends like thy books, few but choice.

Elbert Hubbard

The Broad View

Unlike all the preceding sections of this book, which tend to be pragmatic, this section is somewhat philosophical. The lessons in this section explore concepts and values related to the engineering profession. Topics discussed include the challenge of effecting change, giving back to our profession and community, the primary drivers of engineers versus scientists, the different professional cultures, and looking ahead at our individual futures and the future of our profession.

AH HA! A Process for Effecting Change

Faced with the choice between changing one's mind and proving that there is no need to do so, almost everybody gets busy on the proof.
John Kenneth Galbraith

A client of mine once said, "The only one who enjoys change is a baby with a soiled diaper." Why do we often react negatively to the possibility of change? Why are we so willing to invest heavily in justifying the status quo?

Our frequent knee-jerk resistance to change usually is not, in my view, based on satisfaction with current conditions. Intellectually, we know that just about anything could be improved. Our resistance is more likely to be emotional: we fear giving up that which is known, familiar, and comfortable in exchange for the possibility, but not certainty, of improvement. It's not so much the proposed improvement that frightens us as it is the transition from here to there, the letting go of the familiar before grabbing onto that which could be better.

Growth demands a temporary surrender of security.
Gail Sheehy

Niccolo Machiavelli, the Italian politician and writer, force-fully characterized the intensity of negative reactions to possible change and the challenge of effecting substantive change (Bergin, 1947): "There is nothing more difficult to plan, more doubtful of success, nor more dangerous to manage than the creation of a new system. For the initiator has the enmity of all who would profit by the preservation of the old institutions and merely lukewarm defenders in those who would gain by the new one."

Doing Our Homework: Why Do We Do What We Do?

Prior to starting the change process, we must do our homework. For example, we must ask, "Why do we do it the way we do it?" Maybe the original rationale no longer applies. The following true story emphasizes the importance of determining the origin of a current system.

I once served as dean of an engineering college and, early in my tenure, I noticed that engineering faculty were teaching the physics course to the engineering students. By this time, the dedication and competence of the physics faculty in the college of arts and sciences had become apparent to me. Because of this disconnect, I asked the engineering college department heads why engineering faculty, instead of physics faculty, were teaching physics to engineering students. The answer: the physics faculty developed a reputation for treating engineering students unfairly. That sounded like a good reason and I was tempted to let it lie. However, I asked one more question: When did this happen? Answer: More than 20 years ago! I did some research and found that physics faculty members who were allegedly unfair to engineering students were all either retired or deceased. Soon thereafter, physics was taught to engineering students by physics faculty.

> *All truth goes through three stages.*
> *First it is ridiculed, then it is violently opposed, finally it is accepted as self-evident.*
> Arthur Schopenhauer

Homework also includes identifying stakeholders, that is, individuals, groups, and public, private, and other entities affected by the current and/or the proposed system. Try to anticipate the perceived and actual costs and benefits for each stakeholder and be prepared to acknowledge them. Confirm that the overall "benefits" exceed the overall "costs." This determination typically defies quantification and is likely to require considerable vision and intuition.

Effecting the Change

We "have our ducks in a row." Our homework is completed, and we realize that we may occasionally have to return to the drawing board for refinements. We are ready to advocate a major change within our firm, government entity, academic institution, professional society, or community group. How should we proceed? Does a sure-fire model for change exist? No!

However, one approach that has proven useful is the Awareness-Understanding-Commitment-Action model. It also can be thought of as the Awareness-Head-Heart-Action (AH HA!) approach. The usefulness of this simple, four-step model lies in its focus on helping to understand human behavior in an environment of change and using that understanding to develop a strategy and tactics to effect change.

> *Carry the battle to them. Don't let them bring it to you.*
> Harry S Truman

Awareness. Begin by making stakeholders generally *aware* of a possible major change. Go slow and expect widespread disinterest, suspicion, criticism, skepticism, and resistance, along with some scattered excitement. Explain why the change is needed. Discuss how the contemplated change would, on the balance, benefit the stakeholders. Don't be discouraged. As Machiavelli advised, enmity is expected.

Understanding. Machiavelli also referred to lukewarm support. During the awareness step, look for sources of lukewarm support and focus on those individuals and entities. We should do all we can—ask, listen, talk, meet, write, interact—to help the lukewarm few **understand** why we need the proposed dramatic change and how it will benefit stakeholders. Listen carefully, eliminate semantics hurdles, and refine the proposed change in response to thoughtful concerns and suggestions.

Commitment. Have confidence that some lukewarm supporters will become red hot. They will become mentally and emotionally involved. Their "heads" and "hearts" will be engaged as they leave behind wary awareness and move through understanding and into commitment. Some will be willing to *commit* their reputations, energy, and creativity by becoming advocates. Urge them to help others understand why the proposed change is needed and how it will benefit stakeholders. This education and interaction effort will start a desirable domino effect, in which the commitment of a few expands into the commitment of many. Numerous "heads" and "hearts" will now be on board.

> *Never doubt that a small group of committed people can change the world. It is the only thing that ever has.*
> Margaret Mead

Action. Now ask the ever-expanding committed individuals and organizations to go further. Encourage each to *act,* to take at least one step or perform at least one implementation task. Ask each committed individual and organization to place one building block in what is now becoming the foundation of the change superstructure. Step back and quietly and thankfully watch that superstructure rise as more and more stakeholders become aware, achieve understanding, commit, and act.

While a sharp distinction exists between awareness and understanding and also between understanding and commitment, the differentiation between commitment and action may not be as apparent. The commitment and action steps may appear similar, if not the same. In the absence of action, we may doubt one's commitment.

In response to this concern, there are at least two ways in which bona fide commitment could exist without action. First, while the person may be committed to the change, including the need for it and the benefit of it, he or she may not be familiar enough with the change leaders and their strategies and tactics to be able to identify appropriate acts or tasks. Other individuals may be somewhat introverted (the majority of engineers are) and, therefore, somewhat reluctant to step forward and volunteer their services. Learn to recognize these two types of individuals, invite them to act, and provide them a task to act on. Offer and then provide support. Be confident that essentially all committed individuals will act if invited to do so and asked to contribute in specific ways.

> *Destiny is not a matter of chance, It is a matter of choice.*
> William Jennings Bryan

Widespread Applicability

The effectiveness of the four-step process is essentially independent of the type of organization within which it is applied. The Awareness-Understanding-Commitment-Action approach will work in the business community, government entities, academia, professional societies, and the non-profit sector. The AH HA! model has broad applicability because it is not highly specific and because it recognizes fundamental human behavior, especially the need to engage head and heart.

Suggestions for Applying Ideas

Use the following rules of change to guide your next change effort (Ganz, 2001)

> *Be the change you want to see in the world.*
> Mahatma Gandhi

▶ People can change, but you can't change them. They can only change themselves.

▶ You can only change yourself.

▶ When you change, those you interact with have a new experience, increasing the possibility of change for them.

▶ If you don't change, the power to change belongs to others.

▶ People experience change differently; for some it is exciting, for others it hurts.

▶ Organizations, systems, and individuals typically resist change. Lack of change and/or flexibility leads to extinction.

▶ Identify what will not change, define what will, and acknowledge loss.

▶ Do what you did and you will get what you got.

▶ Change tends to create conflict, sometimes intense conflict. Poorly managed change creates unnecessary conflict.

Because dramatic change is typically proposed to respond to an ominous threat or to seize an unusual opportunity, respond to strong opponents to your proposed change in one of the following ways

▶ Do you agree we are threatened? If so, what change do you propose to meet the threat?

▶ Do you agree we have encountered an opportunity? If so, how do you propose we seize it, or what equivalent or better opportunity do you envision and how do you propose that we pursue it?

> *A jackass can kick down a barn, but it takes a craftsman to build one.*
> Sam Rayburn

▶ The point of the preceding two suggestions: If there is agreement on a threat or opportunity, being against a proposed change to meet the threat or seize the opportunity is not enough. Having an alternative approach is expected.

Recognize that this lesson addresses ways to effect dramatic change, not continual assessment and improvement

▶ Organizational continual assessment and improvement efforts have many other names, including continuous quality improvement and total quality management (TQM) (Walesh, 2000).

> *How wonderful it is that nobody need wait a single moment before starting to improve the world.*
>
> Anne Frank

▶ Both types of change are essential in the lives of individuals and organizations. Thriving individuals and organizations, contrasted with those that are just surviving or worse, are adept at effecting both dramatic change when warranted and gradual change through continual assessment and improvement.

▶ Dramatic change requires an intense effort over a short period of time.

▶ The kind of change resulting from assessment and improvement requires a high degree of self and organizational discipline on an ongoing basis. However, because assessment and improvement are highly dependent on individuals, any one of us can begin right now.

Read the following related lessons

▶ Lesson 7, "Courage: Real and Counterfeit"

▶ Lesson 9, "Go Out on A Limb"

▶ Lesson 13, "Afraid of Dying, or Not Having Lived?"

▶ Lesson 52, "Looking Ahead: Can You Spare a Paradigm?"

Study one or more of the following sources cited in this lesson

▶ Bergin, T.G., ed. and trans. 1947. *Niccolo Machiavelli: The Prince*. Arlington Heights, Ill.: Crofts Classics.

▶ Ganz, J. 2001. "Master the keys to leadership, unlock opportunities." *Engineering Times* (Oct.), p. 6. The rules are attributed to McArdle Ramerman, Inc., a leadership coaching firm.

▶ Walesh, S.G. 2000. *Engineering your future: the non-technical side of professional practice in engineering and other technical fields*, 2nd Ed. Reston, Va.: ASCE Press. Chapter 6, "Total Quality Management," provides an introduction to total quality management and similar continuous assessment and improvement processes.

Refer to the following supplemental source

▶ Kriegel, R., and D. Brandt. 1996. *Sacred cows make the best burgers: paradigm-busting strategies for developing change-ready people and organizations*. New York: Warner Books.

Humorously outlines an organizational change readiness model that consists of these five steps: rounding up sacred cows, developing a change-ready environment, turning resistance into readiness, motivating people to change, and developing the seven personal change-ready traits.

Visit one or more of these websites

▶ "Management First" (http://managementfirst.com/) is a website offered by a British firm. Included is a change management page, which offers free change-related articles.

▶ "Organisational Change" (http://organisationalchange.co.uk/) is the website of a British firm offering change services such as change management training, organizational change consulting, and TQM training. Articles about change are provided free.

In a time of drastic change, it is the learners who inherit the future. The learned usually find themselves equipped to live in a world that no longer exists.

Eric Hoffer

49

Giving to Our Profession and Our Community

> *We only really find ourselves when we lose ourselves in something beyond ourselves.*
>
> Charles Handy

Our Profession

Active involvement in professional organizations is one way to continuously increase the value of personal professional equity. The process of giving something tangible back to the engineering profession helps hone management and leadership ability. Two other effective mechanisms for enhancing personal professional equity are undertaking varied and challenging work assignments and continuing education. However, the first half of this lesson focuses on meaningful involvement in professional organizations.

> *I hold every man a debtor to his profession; from that which man has a course to seek countenance and profit, so ought they of duty to endeavor themselves, by way of amends, to be a help and ornament there unto.*
>
> Francis Bacon

Besides the somewhat selfish concern of maintaining personal professional assets, many of us realize we derive a satisfying and prosperous living from our professions and, accordingly, ought to give something back to them. As practicing engineers, we use the work of many predecessor professionals, many of whom produced the books, papers, conference proceedings, manuals of practice, computer software, and other valuable contributions for which they received little or no monetary compensation.

As an indication of the contribution of others to the profession and our corresponding obligation to do our share, examine your library or the personal library of another successful engineer. Note the relatively large number of materials that were clearly produced, usually in the context of professional organizations, by largely volunteer labor.

The call to be actively involved in professional organizations goes beyond the benefit of maintaining our currency and the need to meet our obligations. Participation provides an opportunity to enjoy and benefit from the company of leaders. The engineering professions and their various subdivisions are like local congregations, mosques, or synagogues—many members, very few doers. The doers are usually committed, creative, ambitious, and accomplished people. All of us, but especially the younger professionals, can learn much from associating with leaders. The "ticket" to that club is a commitment to being actively involved in the work of professional organizations.

> *The world is run by those who show up.*
> Richard Weingardt

Once a commitment is made, we face the challenge of identifying the appropriate one or perhaps several technical and professional organizations that we wish to join. Upon becoming a member of such an organization, select one or more types of activities for involvement and contribution. Besides attending meetings, consider presenting and publishing papers, serving on and chairing technical and non-technical committees, helping to arrange and run meetings and conferences, and serving as an officer.

By investing our time and talent in one or more professional organizations, we will realize the significant return on our investment in terms of knowledge gained; satisfaction of contribution, that is, giving something back; and the association with leaders of our profession.

Our Community

Volunteer efforts, usually quietly and reliably offered, enable many neighborhood, religious, and community organizations to achieve their objectives. How about offering our hard-earned management and leadership skills to assist these organizations?

Many of us know how to make things happen with people in our work environment. We have led and managed teams, departments, offices, projects, and programs. Our experience typically includes various forms of spoken and written communication, often sharpened in the challenging and contentious public arena. We have acquired knowledge, skills, and attitudes that are directly applicable to and highly valued outside of the workplace. We are well-suited for volunteer efforts by virtue of traits such as our orientation toward action, our ability to comprehend the entire system, and our knowledge of infrastructure and the environment.

> *The good news is that a civil engineer was elected to the lieutenant governorship in the state of Texas. The bad news is that this really is news.*
> Bill Ratliff

Numerous and varied opportunities exist within our neighborhood-religious-community setting to contribute our time and apply our management talents. Examples are participating in an American Cancer Society fundraiser, serving on an appointed community committee or board, assisting with church or synagogue fund drives, coaching Special Olympics athletes, and running for elective office.

Some engineers lament, Rodney Dangerfield style, that "we get no respect." Certainly, more meaningful visibility in the community by engineers would contribute to earning more respect. In a more positive vein, by actively participating in neighborhood, religious, and community groups, we can exert influence and make good things happen. Let's share our hard-earned management and leadership knowledge, skills, and attitudes to improve our neighborhoods, religious organizations, and communities.

> *We make a living by what we get, we make a life by what we give.*
> Winston Churchill

Suggestions for Applying Ideas

Build a personal development laboratory (Berson, 2002)

► Participate in volunteer professional and community organizations as a way to experiment with your management and leadership strategies and tactics.
  • This is win-win.
  • By stepping into managing and leading roles in volunteer organizations, you incur less economic and professional risk than you would in your employer's organization.
  • Even if the results are less than you hoped for, the volunteer professional or community organization is likely to benefit from your efforts.

Read the following related lessons

► Lesson 2, "Roles—Then Goals"

► Lesson 9, "Go Out on a Limb"

► Lesson 47, "Eagles and Turkeys"

Study the following source cited in this lesson

► Berson, B. 2002. "PEPP Talk" (Oct.). ["PEPP (Professional Engineers in Professional Practice) Talk" is the e-newsletter of the National Society of Professional Engineers. Bernie Berson, editor of the newsletter, provided the basic laboratory idea.]

Refer to one or more of the following supplemental sources

► ASCE, Government Relations Staff. 2002. "Bridging the gap: engineers and politics." *Leadership and Management in Engineering* (July), pp. 35-37.

Argues that "the shift in power away from Washington and toward the states makes it more important than ever for engineers to take the plunge," that is, proactively influence policy. This is another way of giving.

▶ Ratliff, B. 2002. "Making things better." *Leadership and Management in Engineering* (July), pp. 20-23.

Asserts that engineers "have a tremendous amount to offer to the decision-makers in various legislative bodies, be it in your city council, your county government, your water district, your flood control district, your state, or your nation." The author, an engineer–politician, reminds engineers that having something to offer is not the same as being invited to offer it. "You've got to insert yourself into the process."

*A man there was, and they called him mad;
the more he gave, the more he had.*

John Bunyan

F = ma *and*
You Can't Push a Rope

> *It is the process of design, in which diverse parts of the given-world of the scientist and the made-world of the engineer are reformed and assembled into something the likes of which Nature had not dreamed, that divorces engineering from science and marries it to art.*
>
> Henry Petroski

*T*he title of this lesson is advice given by a faculty member during my sophomore year in engineering school. According to the professor, his words of wisdom were the key to success in engineering practice. The advice made sense, at least superficially, at the time. Now, four decades later, "$F = ma$ and you can't push a rope" is loaded with significance for me, as an engineer and a businessperson. Possibly it has meaning for you. Consider two interpretations of the advice.

Theory plus Common Sense

First, by nature, engineering is a two-sided coin and each side is essential. One side is theory, the other is practical, common-sense application. I think of the stormwater/flood control management projects I've worked on. Scientific principles were meshed with pragmatic considerations to result in constructed, functioning, and personally satisfying facilities.

For example, conservation of mass and energy, elegantly embodied in hydrologic–hydraulic computer models, were combined with nitty-gritty resolution of financial, regulatory, constructability, maintainability, and stakeholder challenges in the process of going from problem to solution, from concept through construction. These projects, and hopefully all or most members of the project team, had to value and use both science ($F = ma$) and pragmatic, common-sense considerations (you can't push a rope). Both sides of the coin were necessary.

Failure to respect science—one side of the coin—can lead to failure. Witness the catastrophic collapse in 1981 of walkways in the Kansas City, Missouri, Hyatt Hotel. Lack of attention to principles of static mechanics was a primary cause of 114 deaths and 200 injuries (Walesh, 2000). Similarly, failure to account for fundamental aerodynamic instability caused the dramatic failure of Galloping Gertie, the Tacoma Narrows Bridge in Washington, in 1940 (Petroski, 1985).

> *It is not a paradox to say that in our most theoretical moods we may be nearest to our most practical applications.*
>
> Alfred North Whitehead

Creating What Never Has Been

The second interpretation of "$F = ma$ and you can't push a rope" is that it illuminates the fundamental difference between engineers and scientists. Experience in both practice and academia has shown me that the gut-level motivation of engineers and scientists is markedly different. Scientists are driven by the desire to understand the physical world about them. This is the source of their professional satisfaction.

Engineers, while valuing and using science, are driven by the desire to do something with that world; to make it better, to directly improve the quality of life. This is the origin of their professional satisfaction. On learning of a scientific discovery, we engineers are likely to say "Congratulations!" to the scientists and then immediately ask "What's it good for; what's the application?" We engineers appreciate science as represented by "$F = ma$," but we derive our satisfaction from its practical application, as suggested by "you can't push a rope."

> *Scientists define what is, engineers create what never has been.*
>
> Theodore von Karman

Engineer and author Samuel C. Florman (1976) argues that talents and impulses deep within us underlie our engineered creation. According to him, our main goal is to "understand the stuff of the universe, to consider problems based on human needs, to propose solutions, . . . and to follow through to a finished product." Creating useful things for society's welfare gives us "existential delight."

Engineering professor and author Hardy Cross probably would have appreciated the advice "$F = ma$ and you can't push a rope." That's what he seemed to be stating, in much more eloquent fashion, when he wrote this about engineering (1952):

> It is not very important whether engineering is called a craft, a profession, or an art; under any name this study of man's needs and of God's gifts that may be brought together is broad enough for a lifetime.

Suggestions for Applying Ideas

Appreciate the constancy of theory and other fundamentals, but be original in their engineering and other applications

▶ Consider the following ways to stimulate creative thought in application of fundamentals:

- Drill down through problem symptoms to find the cause of a problem. While a frequently flooded roadway may be a "problem" to local drivers, to the engineer it is a symptom of a problem. The cause of the problem must be determined before fundamentals and common sense can be creatively applied to solve the problem.
- Read widely, travel to new places, interact with a variety of people, become aware of and learn about new opportunities, problems, and solutions. For example, European travel increased my awareness of the need for river and floodplain restoration and knowledge of how to apply fundamentals to do it.
- Proactively seek opportunities to work on a wide variety of projects and with diverse individuals. View the project–people arena as a stimulating learning opportunity. Note how theories inherent in various disciplines provide the foundation of interdisciplinary work. Note also the creative synergy resulting from the interaction of experts from diverse disciplines.

Read the following related lessons

▶ Lesson 51, "The Two Cultures: Bridging the Gap"

▶ Lesson 52, "Looking Ahead: Can You Spare a Paradigm?"

Study the following sources cited in this lesson

▶ Cross, H. 1952. *Engineers and ivory towers*. New York: McGraw-Hill.

▶ Florman, S.C. 1976. *The existential pleasures of engineering*. New York: St. Martin's Press.

▶ Petroski, H. 1985. *To engineer is human: the role of failure in successful design*. New York: St. Martin's Press.

▶ Walesh, S.G. 2000. *Engineering your future: the non-technical side of professional practice in engineering and technical fields*, 2nd Ed. Reston, Va.: ASCE Press. See Chapter 11, "Legal Framework."

Common sense is genius dressed in working clothes.

Ralph Waldo Emerson

51

The Two Cultures: Bridging The Gap

I believe the intellectual life of the whole of western society is increasingly being split into two polar groups: literary and scientific.

C.P. Snow

*T*hroughout my career, I moved back and forth between the practice world and the academic world. While in campus settings, I found that the liberal arts and the professional programs still constitute two rather distinct groups. These two cultures are alive and well and part of the stimulating diversity we find within and outside of universities. There are still at least two campus camps at most universities.

The Two Cultures

In a 1959 lecture, British scientist and author C.P. Snow (1965) described the "literary culture" and the "scientific culture." Applying Snow's two culture models to many of today's universities, the scientific culture might be defined as what we call professional programs—for example, business, engineering, technology, medicine, and law. The literary culture might be defined as the humanities and social sciences programs.

In describing the gulf between the two cultures, Snow noted that the literary culture views the scientific culture as being "brash and boastful" and "shallowly optimistic, unaware of man's condition." In contrast, the scientific culture views the literary culture as lacking foresight and not being really concerned with the natural world and our standard of living.

Also according to Snow, members of the literary culture see a social, economic, environmental, or other problem as something to dissect, discuss, and debate, but not solve. In contrast, scientific culture members see problems as something to solve. Members of the scientific culture tend to share a systematic, rigorous approach to problem definition and solution. In contrast, the literary culture operates in a less disciplined mode. Snow notes that the literary culture reads more and reads widely. The scientific culture reads less and reads narrowly. And so on.

Herbert Hoover—author, humanitarian, 31st U.S. President, and engineer—spoke for the scientific, or engineering and technical, culture, when he wrote the following (Frederich, 1989):

> It is a great profession. There is the fascination of watching a figment of the imagination emerge through the aid of science to a plan on paper. Then it brings jobs and homes to men. Then it elevates the standards of living and adds to the comforts of life. That is the engineer's high privilege.

Likewise, engineer and author Samuel C. Florman called for the members of the scientific culture to recognize the value of the literary culture when he wrote the following (1987):

> It seems to me that anyone who would call himself college educated—particularly anyone who would call himself a professional—should spend some time in close communion with the great souls, the great thinkers, the great artists, of our civilization. Most particularly, those engineers who would be leaders, those who would participate in the important communal debates, should be acquainted with the thoughts, theories, and philosophies that constitute the foundations of our culture.

What We Share

Let's recognize what we share—the bridges that join us—and use our common values as the basis for working together to make a better world. Rather than dwell on differences, balance dictates that we look for shared interests. It seems to me that there are substantive bridges crossing the gap separating the two cultures.

For example, both cultures value communication. How often we use the trite expression "It was a communication problem" and really mean it! Curricula across some college and university campuses seek to come to grips with the communication challenge by stressing the importance of writing, speaking, listening, and the use of graphics and mathematics.

I believe we also share the thrill of creative work. As suggested by engineering professor and author Henry Petroski (1985), the image of the writer transfixed over a blank sheet of paper next to a wastebasket overflowing with false starts applies equally to the composer beginning a score, the artist starting a painting, the engineer initiating a design, the scientist conceiving a hypothesis, and the business manager formulating a marketing strategy.

In my own work as a civil engineer, my highest satisfactions include involvement in conceiving, planning, designing, and implementing projects; seeing and feeling them come to fruition in dirt and concrete and steel; and being appreciated and valued by others for their form and function. I suspect the exhilaration of satisfying my creative compulsion via engineered works is very similar to the exhilaration felt by the writer, composer, and artist.

Our world continues to shrink. The business and professional community increasingly participates in world markets. Intellectual interests aside, the scientific culture must be increasingly prepared to understand the history, culture, and language of other nations. The literary culture offers many diverse resources to enable that understanding. According to the President of Bard College, Leon Botstein:

> Contrary to what today's focus on high technology might imply, the humanities are more important than ever. Subjects like philosophy, history and literature teach you how to interpret information and how to argue a point of view. . . . Not only the written arts but also music and the visual arts will become increasingly important. Music, for example, teaches valuable lessons about time and space. Similarly, visual thinking is critical to using computers and to manipulating images across multiple dimensions.

Besides getting smaller, our world is using technology to be more productive. The literary culture must acquire basic scientific and technical literacy or risk being out of touch and outside of the sphere of influence. A literary culture that is scientifically and technically illiterate does not bode well for the future. The scientific culture stands ready to assist.

The Builders

Having argued for bridge-building between what C.P. Snow called the literary and scientific cultures, and recognizing that engineering is part of the scientific culture, some thoughts on engineering's diversity and commonality are in order.

Breadth and diversity characterize the engineering profession. Consider, for example, the many and varied engineering education programs accredited by the Accreditation Board for Engineering and Technology (ABET). There are 24 groups of engineering programs, and they run the gamut from aerospace engineering to surveying engineering (ABET, 2003).

Reviewing the works of engineering is another way to demonstrate the breadth and diversity of this profession. Regardless of where and when you are reading this lesson, engineered works are probably evident. As you read, can you see or are you using any of the following: water supply and wastewater systems; highways, railroads, airports, waterways, and all the vehicles that use them; dams, levees, and other flood control structures; electric power generation and distribution systems; medical buildings and equipment; and bridges and other large structures? Engineers are known by the creativity and usefulness of their works. The preceding list just scratches the surface.

A third way of illustrating the breadth and diversity of engineering is to reflect on the spectrum of functions carried out by engineers. Included are planning, design, construction, operations, research and development, teaching, marketing, management, and leadership. Because of the range of functions, and thus the range of interests and talents needed to carry out those functions, there is "room" in engineering for a wide variety of individuals.

Engineers also differ markedly in their views, which further enriches the profession's breadth and diversity. Witness the strong feelings expressed on issues such as the engineer's image, theory versus application, the computer's role, length and content of formal education, and licensure.

Throughout engineering's long history and within its great breadth and diversity, however, there is at least one widely shared interest and function: building. In the final analysis, when everything else is stripped away, the engineer is at the core a builder, and building is the glue that binds engineers together. When civil engineers "build," they usually call the process construction. When mechanical engineers "build," they routinely refer to it as manufacturing. From the perspective of electrical engineers, "building" is often referred to as fabrication.

Whatever you call the process, the ultimate end of the engineering process is to "build" something that never before existed and that will meet human needs. Examples include the water supply system "built" by the civil engineer, the energy-efficient automobile "built" by the mechanical engineer, and the electrical power distribution system "built" by the electrical engineer. Some engineers "build" less visible but nevertheless important entities such as computer programs, better ways to perform engineering functions, and improved ways to organize engineering organizations.

I often ask student and practicing engineers, "Why engineering?" Regardless of their age, building and creating very often are part of the answer. As a child I walked across the highway to the Lake Michigan shore to build. I "built" dams, levees, and channels at a point where a creek flowed into the lake. Although old enough to "build," I wasn't old enough to cross the highway, so my mother took me. Later, on learning that civil engineers build water control structures, I decided to study, practice, and earn my living in the water resources area of civil engineering. I now look, with quiet pride, at water control structures and other useful things that I helped to engineer. Most engineers have similar stories in which the common theme is building, or some variation on it.

> *We recognize that we cannot survive on meditation, poems and sunsets. We are restless. We have an irresistible urge to dip our hands into the stuff of the earth and do something with it.*
>
> Samuel C. Florman

Interestingly, when laypersons—who, surveys suggest, know little about engineering—want to convey the idea that something noteworthy has been built, the name of our profession is often invoked. Some politician, business executive, or religious leader is said to have "engineered" this or that.

As is often the case with breadth and diversity, engineering's strength is its breadth and diversity. It enables our profession to cover all the bases and build for the common good. As Herbert Hoover said, "To the engineer falls the job of clothing the bare bones of science with life, comfort, and hope." Building for that purpose is indeed a high calling.

Suggestions for Applying Ideas

Study one or more of the following sources cited in this lesson

▶ ABET. October 6, 2003. www.abet.org. ABET's vision is to "provide world leadership to assure quality and stimulate innovation in engineering, technology and applied science education."

▶ Florman, S.C. 1987. *The civilized engineer.* New York: St. Martin's Press.

▶ Fredrich, A.J. (ed.). 1989. *Sons of Martha: civil engineering readings in modern literature.* New York: ASCE.

▶ Petroski, H. 1985. *To engineer is human: the role of failure in successful designs.* New York: St. Martin's Press. (Stresses the design aspect of the process that ends with building. Advises the engineer that each design is a hypothesis that is not fully tested until it is constructed, manufactured, fabricated, or otherwise built. Advocates the study of failures.)

▶ Snow, C.P. 1965. *The two cultures: a second look.* Cambridge: Cambridge University Press.

Refer to one or more of the following supplemental sources

▶ ASCE, Task Committee on the First Professional Degree. 2001. *Engineering the future of civil engineering* (Oct.). Reston, Va.: ASCE. The report is available at www.asce.org/raisethebar.

Presents a plan for full realization of ASCE's Policy Statement 465, which states that the Society "supports the concept of the Masters degree or equivalent as a prerequisite for licensure and the practice of civil engineering at the professional level." The policy calls for a deeper and broader education for civil engineers.

▶ ASCE, Body of Knowledge Committee of the Task Committee on Academic Prerequisites for Professional Practice. 2003. *Civil engineering body of knowledge for the 21st century: preparing the civil engineer for the future.* Reston, Va.: ASCE. The report is available at www.asce.org/raisethebar.

Describes the broader and deeper body of knowledge (BOK), that is, knowledge, skills, and attitudes that will be needed by future civil engineers to enter professional practice. The BOK is described in terms of its three dimensions: what should be taught and learned, how it should be taught and learned, and who should teach and learn it.

▶ Cross, H. 1952. *Engineers and ivory towers.* New York: McGraw-Hill.

Sharply distinguishes between science and engineering while respecting the value of each. Claims that engineers are more humanists than scientists because they plan, design and build to meet human needs and, in the process, grapple with technology, law, economics and sociology. According to Cross, "The glory of the adaptation of science to human needs is that of engineering." Young engineers, especially those in or contemplating advanced formal education, should read this old book with timeless messages.

Once a mind is truly stretched,
it never returns to its original dimensions.
Anonymous

Looking Ahead:
Can You Spare a Paradigm?

> *Sacred cows make the best burgers.*
> Robert Kriegel and David Brandt

*A*s individuals and as organizations, we should be preparing ourselves for the way our profession will be practiced, not the way it is or was practiced. Looking well into the early 21st century, the workforce will be increasingly heterogeneous, with greater participation by women, ethnic minorities, and seniors. Future engineers and other technical professionals will increasingly serve people and organizations around the globe. Finally, the logistics of work will change with movement toward teleworking and independent contractors; a more varied education-work-retirement pattern; the continued loss of job security; and, for the proactive, attainment of career security. However, our basic mission will not change. We will continue to focus on meeting society's basic physical needs (Walesh, 2000).

Anticipating the future is very difficult. An important quality is being able to "take off the blinders," which naturally leads to the topic of paradigms. Speaker and author Stephen Covey (1990) defines a paradigm as "The way we 'see' the world—not in terms of our visual sense of sight, but in terms of perceiving, understanding, interpreting." According to futurist Joel Barker (1989), a paradigm is "a set of rules and regulations that defines boundaries and tells you what to do to be successful within those boundaries." Psychologists Robert Kriegel and David Brandt (1996) suggest that a paradigm is like the sandbox you played in as a child—it was your world.

Paradigms abound; they are all around us, as seen from the following examples:
- The traditional, pyramidal organizational structure of consulting organizations, construction firms, manufacturing operations, and academic institutions.
- The nine-month school year.

- Men and women participating in high school athletics.
- Japanese products being of high quality, the implication being that manufacturing organizations hoping to compete globally must match or exceed the quality of Japanese products.
- The 40-hour work week.
- The largely individual-based, competitive learning situation utilized in U.S. higher education.

Paradigms are very useful, given the complexity of society. They almost always allow for more than one "right" answer. For example, the nine-month school year is generally recognized as the norm but is applied in various ways. In some situations, the nine-month school year paradigm consists of two semesters, and in others, it consists of three quarter sessions. Year-round school would be a new paradigm.

> *We do not see things as they are. We see things as we are.*
> H. Jackson Brown, Jr.

On the negative side, Joel Barker claims that paradigms tend to reverse the "seeing and believing" process (1989). Intelligent, thoughtful people like to think that they are rational, that they "believe because they see." However, because of paradigms, people often "see because they believe." Consider, for example, some beliefs you hold and have held for a long time and highly value. Are you not likely to find many examples of situations which tend to support your belief, and might you not actually be looking for them? And might you be "seeing because you believe" rather than "believing because you see?"

Unfortunately, if paradigms are too strongly held, and they often are, the holders risk incurring paradigm paralysis. Paradigm pliancy is a much better strategy, according to Barker, especially in turbulent times. Pliancy is a quality or state of yielding or changing. Fortunately, at least a handful of people in any organization can change their paradigms. Even for them, paradigm pliancy is at best difficult.

As suggested by progression from the abacus to the slide rule to the electronic calculator to the digital computer, familiar to older engineers, paradigms do shift. Many examples can be found within technical professions and throughout society at large. Paradigms vanish and new ones replace them.

> *You can never plan the future by the past.*
> Edmund Burke

What does the future hold? The most successful individuals and organizations will avoid paradigm paralysis—they will practice paradigm pliancy and they will thrive. They will create new paradigms and build bridges to them, or at least they will recognize new paradigms when they are coming down the pike and see the business and professional opportunities within them. In contrast, paradigm paralysis characterizes individuals and organizations satisfied with surviving or in the process of dying. What paradigms will you contribute to your engineering or other technical field, or what paradigms created by others will you enthusiastically embrace and advance?

When all is said and done, there are only two futures for individuals and organizations. The first is the future we create for ourselves. The second is the future others create for us. If we, as individuals or organizations, don't choose the first, others will impose the second.

Suggestions for Applying Ideas

Identify possible paradigm shifts by comparing situations that existed several decades ago in the U.S. to those that exist today

- Inter-high school and college sports were essentially exclusively for males (Barker, 1989).
- The slide rule was the personal computer, and a digital computer with its peripherals typically filled an entire room. Ken Olsen, former president of computer manufacturer Digital Electronic Corporation, once said, "There is no reason for any individual to have a computer in their home" (Barker, 1989).
- There was a strong feeling that nuclear power would soon solve most energy problems throughout the U.S. and that many nuclear power plants would be built in successive decades (Barker, 1989).
- Watches were mechanical and included complex gear and spring mechanisms.
- Products manufactured in Japan were inferior to U.S. products.
- Mail routinely took days to be delivered.
- The worldwide conflicts between democracy and communism and capitalism and socialism would continue forever.
- Women did not study engineering or become doctors.

▶ All of the preceding paradigms have vanished and been replaced by new paradigms.

Consider some of the paradigm shifts of the last few decades, with emphasis on the "shifters," "outsiders," and "oddballs" responsible for them

▶ Fred Smith, founder of Federal Express, wrote a paper while he was a Yale University student that proposed overnight mail delivery in the U.S. using trucks and airplanes operating within a hub and spoke system. According to the professor who gave Smith a "C" on the paper, the idea was interesting, but would never work (Barker, 1989; Cypert, 1993).

> *The empires of the future are the empires of the mind.*
> *Winston Churchill*

▶ Chester Carlson developed the process of electrostatic photography. He offered it to 43 companies in the late 1940s but found only one organization with foresight, resulting in the development of what is now called xerography. The problem at the time was that the photography paradigm consisted of film, developer, and a darkroom—there was no other way to do it. At least 43 companies could not envision the now omnipresent copy machines (Barker, 1989).

▶ High-jumper Dick Fosbury was ridiculed for leading with his head at a time when all others jumped feet first. However, the ridiculed "Fosbury flop" enabled Fosbury to win the high-jump gold medal in 1968 at Mexico City. Leading with the head is now the standard for world-class high-jumpers (Kriegel and Brandt, 1996).

▶ Swiss watch researchers in Neuchatel, Switzerland, developed a novelty quartz watch with a digital display in 1967. At that time, Swiss mechanical watches enjoyed 60% of the global watch market. They displayed their unprotected novelty watch at an international watchmaker's conference. The idea was picked up by Japanese entrepreneurs and, as a result, the Swiss share of the global watch market fell to less than 10% (Barker, 1989).

> *Do not follow where the path may lead. Go instead where there is no path and leave a trail.*
>
> *Anonymous*

▶ W. Edwards Deming's advice on what is now called total quality management was ignored by American businesses. Under the auspices of the U.S. government, he assisted the Japanese after World War II, who, in a matter of a few decades, set the world standard for manufactured products (Barker, 1989).

Examine, at the most fundamental level, the assumptions, approaches, techniques, and tools you use in your professional work and, perhaps, beyond

The "you" in this advice applies to individuals and organizations. What are you doing simply because you have "always" done it? Driven by your vision, and consistent with your mission and core values, what could and should be cast out and replaced? What might you try to do that you are not doing? Some ideas for individuals and/or organizations are as follows:

▶ Explore possible pockets of paradigm paralysis by obtaining ideas and information from external sources, such as the following (Hayden, 2003):

- New employees within their first few months before they absorb your culture.
- Bright professionals who left your organization at the first opportunity.
- Clients who consistently reject your proposals.
- Clients or customers who "fired" you.

▶ Eliminate that sacred decades-old service line that no longer is profitable.

▶ Give every employee one day off every six months to work as a volunteer somewhere in the community. Think of the good and goodwill to be achieved, the perspectives and knowledge gained, and the contacts and connections made (Boller, 2003).

▶ Cease rewarding personnel for seniority, for just putting in time.

▶ Institute a "walk a mile in my shoes" program to enable personnel to broaden and deepen their knowledge and skills and reexamine their attitudes. Temporarily or permanently shift personnel from one function, position, or organizational unit to another (Boller, 2003).

▶ Stop thinking first of why something can't be done. Think instead of ways to do it.

▶ Revisit your organization's "bigger is better" strategy. Might fewer employees filling crucial or core producing, managing, and leading roles be more productive, more profitable, less stressful, more enjoyable, and otherwise better? This assumes they would be supported by a cadre of independent consultants and support personnel, who participate only when needed and then interact partly via electronic means.

▶ Take ideas to the city council, instead of reacting to theirs.

▶ Think of ways some of your firm's services could be packaged as education and training products and marketed over the web.

▶ Conduct post-audits of representative projects and objectively analyze what you did right and wrong. Invite junior members of the organization to actively participate in the critique and avoid "thin skin." The U.S. Army has used after-action reviews for more than a decade, and their worth has been proven (Galloway, 2003).

▶ Contemplate the idea that your firm might be able to differentiate itself and provide much more value to clients by designing and servicing "smart" structures and facilities. That is, embed sensors and transmitters and then, on a multi-year contract basis, provide the client with structure and facility performance analysis, management advice, and as-needed operation and design services (Russell, 2003).

The overall idea is to occasionally examine, in a fresh and critical manner, the basics of what we are doing. This goes beyond continuous improvement, as important as that is. The goal is to find those aspects of our work life and beyond that could and maybe should be radically changed for our benefit and the benefit of our employers, our clients, and others within our circle of influence.

Refer to the following related lessons

▶ Lesson 9, "Go Out on a Limb"

▶ Lesson 13, "Afraid of Dying, or Not Having Lived?"

▶ Lesson 31, "We Don't Make Whitewalls: Work Smarter, Not Harder"

▶ Lesson 48, "AH HA! A Process for Effecting Change"

Study one or more of the following sources cited in this lesson

▶ Barker, J.A. 1989. *Discovering the future: the business of paradigms*. St. Paul, Minn.: ILI Press.

▶ Boller, R. 2003. High school principal, retired. Personal communication, July 14.

▶ Covey, S.R. 1989. *The 7 habits of highly effective people: restoring the character ethic*. New York: Simon & Schuster, p. 23.

▶ Cypert, S.A. 1993. *The success breakthrough: get what you want from your career, your relationships, and your life*. New York: Avon Books, pp. 201-202.

▶ Galloway, G.E., Jr. 2003. Titan Systems Corporation. Personal communication, July 22.

▶ Hayden, W.M., Jr. 2003. Wendel-Duchscherer Architects and Engineers. Personal communication, July 12.

▶ Kriegel, R., and D. Brandt. 1996. *Sacred cows make the best burgers: paradigm-busting strategies for developing change-ready people and organizations.* New York: Warner Books, p. 56.

▶ Russell, J.S. 2003. University of Wisconsin-Madison. Personal communication, July 15-16.

▶ Walesh, S.G. 2000. *Engineering your future: the non-technical side of professional practice in engineering and other technical fields,* 2nd Ed. Reston, Va.: ASCE Press. See Chapter 15, "The Future and You."

I don't set trends. I just find out what they are and I exploit them.

Dick Clark

Appendix:
Sources of Quotations

Numbers refer to the lesson that contains a quotation by the individual listed. Unless otherwise noted, individuals are from the United States.

Robert F. ABBOTT, journalist, 14
Konrad ADENAUER, Chancellor of the Federal Republic of Germany, 13
AESOP, Greek slave who authored fables, 7
Edward ALBEE, playwright, 20
Thomas Bailey ALDRICH, father of the American novel, 27
James ALLEN, English writer, 32
Mary Ann ALLISON, business author, 45
ARISTOTLE, Greek philosopher, 7
Neil ARMSTRONG, astronaut, 18
Isaac ASIMOV, author, 25
Marcus AURELIUS, Roman philosopher-emperor, 32

Roger W. BABSON, investment banker, 10
Marcus BACH, author, 13, 16
Francis BACON, English philosopher and statesman, 9, 27, 49
Byrd BAGGETT, salesman, 5
Sunny and Kim BAKER, authors and management consultants, 36
Joel A. BARKER, futurist, 52
Roland BARTH, teacher and principal, 30
Harry BECKWITH, author, 42
Henry Ward BEECHER, minister, writer, and speaker, 1
Alexander Graham BELL, inventor, 6, 34
Gary BELSKY, professor and journalist, 10
Warren G. BENNIS, leadership author, 1
Debra A. BENTON, consultant and author, 16
Milton BERLE, comedian, 41
Yogi BERRA, Hall of Fame baseball player, 3, 38
John Shaw BILLINGS, physician and librarian, 19
David P. BILLINGTON, engineering professor, 25
Larry BIRD, basketball player, 3

Bibliography

Books and periodicals

Abbott, R.F. 1999. "Downward communication: enabling communication" (Sept. 1); "Upward communication: compliance" (Sept. 8); "Lateral communication: coordination" (Sept. 22); "The grapevine: defying the rules" (Sep. 29). *Abbott's Communication Letter.*

Advanced Public Speaking Institute. 2003. "Public speaking: stage fright." *Leadership and Management in Engineering* (Jan.). pp. 4-5.

Allen, C. 2001. "Using the power of diversity to retain staff: developing tools to ensure success." *Leadership and Management in Engineering* (Winter), pp. 22-25.

Allen, J. 1983. *As a man thinketh.* Marina Del Ray, Calif.: DeVorss & Company.

Aristotle. 1987. *The Nicomachean ethics.* Trans. by D. Ross and revised by J.L. Ackrill and J.O. Urmson. Oxford: Oxford University Press.

ASCE. 2002. "Code of Ethics," *ASCE Official Register 2002.* Reston, Va.: ASCE.

ASCE, Government Relations Staff. 2002. "Bridging the gap: engineers and politics." *Leadership and Management in Engineering* (July).

ASCE. No date. *Mentors and mentees.* Reston, Va.: ASCE.

Ashton, A., and R. Ashton. 1999. "Long distance relationships." *Home Office Computing* (July), p. 52-55.

Avila, E.A. 2001. "Competitive forces that drive civil engineer recruitment and retention." *Leadership and Management in Engineering* (July), pp. 17-22.

Bach, M. 1970. *The world of serendipity.* Marina Del Ray, Calif.: DeVorss & Company.

Baker, M. III. 1997. "Reengineering an engineering consultant." *Journal of Management in Engineering* (March/April), pp. 20-24.

Baker, S., and K. Baker. 1998. *The complete idiot's guide to project management.* New York: Alpha Books.

Barker, J.A. 1989. *Discovering the future: the business of paradigms.* St. Paul, Minn.: ILI Press.

Bell, C.R. 1996. *Managers as mentors.* San Francisco, Calif.: Koehler Publishers.

Belsky, G., and T. Gilovich. 1999. *Why smart people make big money mistakes—and how to correct them.* New York: Simon & Schuster.

Bennis, W.G. 1989. *Why leaders can't lead—the unconscious conspiracy continues.* San Francisco, Calif.: Jossey-Bass Publishers.

Bennis, W.G., and R.J. Thomas. 2002. *Geeks and geezers: how era, values, and defining moments shape leaders*. Boston, Mass.: Harvard Business School Press.

Benton, D.A. 1994. *Lions don't need to roar: using the leadership power of professional presence to stand out, fit in and move ahead*. New York: Warner Books.

Bergin, T.G., ed. and trans. 1947. *Niccolo Machiavelli: The Prince*. Arlington Heights, Ill.: Crofts Classics.

Berson, B. 2002. "PEPP Talk" (Oct.). Professional Engineers in Professional Practice Newsletter.

Berthouex, P.M. 1996. "Honing the writing skills of engineers." *Journal of Professional Issues in Engineering Education and Practice* (July).

Billington, D.P. 1996. *The innovators: the engineering pioneers who made America modern*. New York: John Wiley & Sons.

Bode, R. 1993. *First you have to row a little boat*. New York: Warner Books.

Bonar, R.L., and S.G. Walesh. 1995. "Ownership transition: a mentoring case study." Proceedings of the Fall Conference of the American Consulting Engineers Council. Washington, D.C.: American Consulting Engineers Council.

———. 1998. "Mentoring: an investment in people." Proceedings of the Fall Conference of the American Consulting Engineers Council. Washington, D.C.: American Consulting Engineers Council.

Bowler, P.L. 1990. *Readable writing handbook*. Helena, Mont.: Montana Dept. of Highways.

Braun, C.F. 1954. *Management and leadership*. Alhambra, Calif.: C.F. Braun & Company.

Brown, T.L. 1986. "Time to diversify your life portfolio?" *Industry Week* (Nov. 10), p. 13.

———. 1992. "Teams can work great." *Industry Week* (Feb. 17), p. 18.

Camp, R.C. 1989. *Benchmarking: the search for industries best practices that lead to superior performance*. Milwaukee, Wis.: ASQ Quality Press.

Carlson, R. 1997. *Don't sweat the small stuff . . . and it's all small stuff: simple things to keep the little things from taking over your life*. New York: Hyperion.

Carnegie, D. 1935. *Public speaking and influencing men in business*. New York: Association Press.

Chopra, D., and D. Simon. 2002. *Grow younger, live longer: 10 steps to reverse aging*. New York: Three Rivers Press.

Cialdini, R.B. 2001. "The science of persuasion." *Scientific American* (Feb.), pp. 76-81.

Coe, J.J. 1987. "Engineers as managers: some do's and don'ts." *Journal of Management in Engineering* (Oct.), pp. 281-287.

Collins, J. 2001. *Good to great: why some companies make the leap and others don't*. New York: HarperCollins.

Commission on the Advancement of Women and Minorities in Science, Engineering and Technology. 2001. "Land of plenty: diversity as America's competitive edge in science, engineering and technology." *Leadership and Management in Engineering* (Oct.), pp. 27-30.

Cori, K.A. 1989. "Project work plan development." Paper presented at the Project Management Institute and Symposium, Atlanta, Ga. (Oct.).

Covey, S.R. 1989. *The 7 habits of highly effective people: restoring the character ethic*. New York: Simon & Schuster.

Creed, M.W. 1999. "Maximum impact: organizing your presentation." *Journal of Management in Engineering* (Sep./Oct.), pp. 28-31.

Cross, H. 1952. *Engineers and ivory towers*. New York: McGraw-Hill.

Culp, G., and A. Smith. 1997. "Six steps to effective delegation." *Journal of Management in Engineering* (Jan./Feb.), pp. 30-31.

Cypert, S.A. 1993. *Success breakthrough: get what you want from your career, your relationships and your life.* New York: Avon Books.

de Camp, L.S. 1963. *The ancient engineers.* New York: Ballantine.

Decker, B., and J. Denney. 1992. *You've got to be believed to be heard.* New York: St. Martin's Press.

Delatte, N.J. 1997. "Failure case studies and ethics in engineering mechanics courses." *Journal of Professional Issues in Engineering Education and Practice* (July), pp. 111-116.

De Pree, M. 1989. *Leadership is an art.* New York: Dell Publishing Company.

DPIC Companies. 1993. *The contract guide: DPIC's risk management handbook for architects and engineers.* DPIC.

Dunham, C.W., R.D. Young, and J.T. Bockrath. 1979. *Contracts, specifications and law for engineers,* 3rd Ed. New York: McGraw-Hill.

Economist. 1999. "When companies connect" (June 26), pp. 19-20.

Elovitz, K.M. 1995. "Avoiding the 10 'demandments' of contract negotiations." *Consulting – Specifying Engineer* (Sep.), pp. 15-19.

Emerson, R.W. 1936. *Essays.* Reading, Pa.: The Spencer Press.

Engineering Times. 1992. "A/E firms hit or miss on training" (Feb.).

ENR. 1988a. "Key beam under-designed" (July 7), p. 14.

———. 1988b. "German bridge girder fails" (Sept. 8), p. 18.

———. 1990a. "Design flaw blamed for collapse" (Jan. 18), pp. 11-12.

———. 1990b. "Design led to downfall of incremental launch" (Feb. 1), pp. 13-14.

Eschenbach, R.C., and T.G. Eschenbach. 1996. "Understanding why stakeholders matter." *Journal of Management in Engineering* (Nov./Dec.), pp. 59-64.

Evans, M.D. 1995. "Student and faculty guide to improved technical writing." *Journal of Professional Issues in Engineering Education and Practice* (April), pp. 114-122.

Fairchild, F.P., and H.E. Freeman. 1993. "Establishing a formal mentoring program in a consulting engineering firm." Presented at the ASCE Engineering Management Conference, Denver, Colo. (Feb.).

Farr, J.V., and J.F. Sullivan, Jr. 1996. "Rethinking training in the 1990s." *Journal of Management in Engineering* (May/June), pp. 29-33.

Farris, G. 1997. "Citizens advisory groups: the pluses, the pitfalls and better options." *Water/Engineering and Management* (Oct.), pp. 28-31.

Fenske, S.M., and T.E. Fenske. 1989. "Business planning for new engineering consulting firms." *Journal of Management in Engineering* (Jan.), pp. 89-95.

Fitzpatrick, B. 1997. *100 action principles of the Shaolin.* Natick, Mass.: American Success Institute.

Florman, S.C. 1976. *The existential pleasures of engineering.* New York: St. Martins Press.

———. 1987. *The civilized engineer.* New York: St. Martin's Press.

Frank, M.O. 1989. *How to run a successful meeting in half the time.* New York: Simon & Schuster.

Fredrich, A.J. (ed.). 1989. *Sons of Martha: civil engineering readings in modern literature.* New York: ASCE.

Freese and Nichols. No date. "Freese and Nichols University: Where Education Is a Tradition."

Galloway, P.D. 2001. "Innovative benefits in a small consulting firm." *Leadership and Management in Engineering* (Winter), p. 45-47.

Ganz, J. 2001. "Master the keys to leadership, unlock opportunities." *Engineering Times* (Oct.), p. 6.

Gerber, R. 2002. *Leadership the Eleanor Roosevelt way: timeless strategies from the first lady of courage.* Upper Saddle River, N.J.: Prentice Hall.

Glagola, C.R.F., and C. Nichols. 2001. "Recruitment and retention of civil engineers in Department of Transportation." *Leadership and Management in Engineering* (Winter), pp. 30-36.

Goodwin, D.K. 1994. *No ordinary time*. New York: Simon & Schuster.

Green, L. 1995. "The 7 habits of highly ineffective people." *American Way* (Aug. 15), pp. 56-60.

Grugal, R. 2002a. "Reading is an art form." *Investor's Business Daily* (May 5).

———. 2002b. "Get past the gatekeepers." *Investor's Business Daily* (Oct. 22).

Hammer, M., and S.A. Stanton. 1995. *The reengineering revolution: a handbook*. New York: HarperCollins.

Handy, C. 1998. *The hungry spirit: beyond capitalism: a quest for purpose in the modern world*. New York: Broadway Books.

Hayden, W.M., Jr. 1987. *Quality by Design Newsletter*. A/E QMA. Jacksonville, Fla. (May).

———. 2002. "Navigating the white water of project management." *Leadership and Management in Engineering* (April), pp. 20-22.

Heery, J.J., Jr. 2001. "Mathematics error: when computation leads to disaster." *Engineering Times* (Oct.).

Hendricks, M. 1995. "More than words." *Entrepreneur* (Aug.), pp. 54-57.

Hensey, M. 1991. "Keys to better meetings." *Civil Engineering* (Feb.), pp. 65-66.

———. 1995. *Continuous excellence: building effective organizations*. Reston, Va.: ASCE Press.

———. 1999. *Personal success strategies: developing your potential*. Reston, Va.: ASCE Press.

———. 2001. "Innovations and best practices: leadership development and retention." *Leadership and Management in Engineering* (Winter), pp. 37-41.

Herrin, J.C., and A.W. Whitlock. 1992. "Interfacing with the public on water-related issues: what TVA is doing." *Saving a Threatened Resource*. New York: ASCE, pp. 293-298.

Hessen, C.N., and B.J. Lewis. 2001. "Steps you can take to hire, keep and inspire generation Xers." *Leadership and Management in Engineering* (Winter), pp. 42-44.

Hill, N. 1960. *Think and grow rich*. New York: Fawcett Crest.

Hirsch, H.L. 2003. *Essential communication strategies for scientists, engineers, and technology professionals*, 2nd Ed. Piscataway, N.J.: IEEE Press.

Hodge, C.S. 1997. "Misunderstanding computer accuracy leads to project rework: two case studies." *Forensic Engineering: Proceedings of First Congress*. Reston, Va.: ASCE.

Hunter, J.C. 1998. *The servant: a simple story about the true essence of leadership*. Rocklin, Calif.: Prima Publishing.

Johnson, H.M., and A. Singh. 1998. "The personality of civil engineers." *Journal of Management in Engineering* (July/Aug.), pp. 45-56.

Kaminski, J.P. 1994. *Citizen Jefferson: the wit and wisdom of an American sage*. Madison, Wis.: Madison House Publishers.

King, R.T. 1996. "The company we don't keep." *The Wall Street Journal* (Nov. 18).

Kiyosaki, R.T., with S.L. Lechter. 1998. *Rich dad, poor dad: what the rich teach their kids about money—that the poor and middle class do not!* New York: Warner Books.

Kotter, J.P. 1999. *John Kotter on what leaders really do*. Cambridge, Mass.: Harvard Business School Press.

Kriegel, R., and D. Brandt. 1996. *Sacred cows make the best burgers: paradigm-busting strategies for developing change-ready people and organizations*. New York: Warner Books.

Kushner, H.S. 2001. *Living a life that matters*. New York: Alfred A. Knopf.

Lantos, P.P. 1998. "Marketing 101: how I got my ten largest assignments." *Journal of Management Consulting* (Nov.), pp. 38-40.

Leeds, D. 2000. *The 7 powers of questions: secrets to successful communication in life and at work*. New York: Perigee.

Leuba, C.J. 1971. *A road to creativity—Arthur Morgan—engineer, educator, administrator*. North Quincy, Mass.: Christopher Publishing House.

Lindeburg, M.R. 1997. *Getting started as a consulting engineer*. Belmont, Calif.: Professional Publications.

Loeffelbein, B. 1992. "Euphemisms at work." *The Rotarian* (Feb.): pp. 22-23.

Lorsch, J.W., and T.J. Tierney. 2002. "Build a life, not a resume." *Consulting to Management* (Sep.), pp. 44-52.

Manchester, F.A., and W.F. Giese (eds.). 1926. *Harper's anthology for college courses in composition and literature*. New York: Harper & Brothers.

Mandino, O. 1968. *The greatest salesman in the world*. New York: Bantam Books.

Martin, J. (pseudonym). 1988. *To rise above principle: the memoirs of an unreconstructed dean*. Urbana, Ill.: University of Illinois Press.

Martin, P., and K. Tate. 1999. "Climbing to performance." *PM Network* (June), p. 14.

Martin, P.K., and K. Tate. 2001a. "Not everything is a project." *PM Network* (May), p. 25.

———. 2001b. "A project management genie appears." *PM Network* (Aug.), p. 18.

Maxwell, J.C., and Z. Zigler. 1998. *The 21 irrefutable laws of leadership*. Nashville, Tenn.: Thomas Nelson Publishers.

McCormack, M.H. 2000. *Staying street smart in the Internet age: what hasn't changed about the way we do business*. New York: Viking Press.

McGregor, D. 1960. *The human side of enterprise*. New York: McGraw-Hill.

Meister, J.C. 1994. *Corporate quality universities*. New York: Irwin Professional Publishing.

Messersmith, J., and A. Schrader. 2003. "Bringing people, processes, and the work place together to create high-performance work environments." Forum, *Leadership and Management in Engineering* (Jan.), pp. 5-7.

Mink, M. 2001. "Inventor Ray Kurzweil: his passion to create helped give blind people their independence." *Investors Business Daily* (July 10).

Mole, J. 1998. *Management mole: lessons from office life*. London: Bantam Press.

Morrell, K., and M. Simonetto. 1999. "Managing retention at DeLoitte Consulting." *Journal of Management Consulting* (May), pp. 55-60.

Murphy, J. 1963. *The power of your subconscious mind*. Englewood Cliffs, N.J.: Prentice Hall.

National Institute of Business Management. 1988. *Body language for business success*. New York: NIBM.

Neville-Scott, K. No date. *Technical writing/communications workshop: taking the mystery out of writing*. Cambridge, Mass.: Camp Dresser & McKee, Inc.

Newton, M. 2001. "Interviewing tips." *Leadership and Management in Engineering* (July), pp. 5-6.

NSPE. 2002. "New study shows structural engineers are still high-risk discipline for liability claims." *Engineering Times* (April).

O'Connell, M. 1996. "Training as a potential profit center." *Journal of Management in Engineering* (Sept./Oct.), pp. 25-27.

O'Dell, C., and C.J. Grayson, Jr. 1998. *If only we knew what we know: the transfer of internal knowledge and best practice*. New York: Free Press.

Oxer, J.P. 1999. "The independent contractor." *Journal of Management in Engineering* (Jan./Feb.), pp. 18-20.

Parkin, J. 1997. "Choosing to lead." *Journal of Management in Engineering* (Jan./Feb.), pp. 62-63.

Peck, M.S. 1997. *The road less traveled and beyond: spiritual growth in an age of anxiety.* New York: Simon & Schuster.

Perlstein, D. 1998. *Solo success: 100 tips for becoming a $100,000-a-year freelancer.* New York: Three Rivers Press.

Peters, T. 1994. *The pursuit of WOW!* New York: Vintage Books.

Petroski, H. 1985. *To engineer is human: the role of failure in successful design.* New York: St. Martin's Press.

———. 1993. "Engineers as writers." *American Scientist* (Sept./Oct.), pp. 419-423.

Pirsig, R.M. 1974. *Zen and the art of motorcycle maintenance.* New York: Bantam Books.

Pitzrick, D.A. 2001. "One company's approach to recruitment and retention." *Leadership and Management in Engineering* (Winter), pp. 48-50.

Project Management Institute. 2000. "A guide to the project management body of knowledge." Newtown Square, Pa.: PMI.

Rad, P.F. 2001. "From the editor." *Project Management Journal* (June), p. 3.

Raskin, A. 2002. "The color of cool." *Business 2.0* (Nov.), pp. 49-52.

Ratliff, B. 2002. "Making things better." *Leadership and Management in Engineering* (July), pp. 20-23.

Reinhold, B.B. 1997. "Body language." *US Airways Magazine* (March).

RoAne, S. 1988. *How to work a room: a guide to successfully managing the mingling.* New York: Shapolsky Publishers.

Roesner, L.A., and S.G. Walesh. 1998. "Corporate university: consulting firm case study." *Journal of Management in Engineering* (March/April), pp. 56-63.

Roger, J., and P. McWilliams. 1991. *Do it! Let's get off our butts.* Los Angeles, Calif.: Prelude Press.

Rosenbluth, H.F., and D.M. Peters. 2002. *The customer comes second.* New York: Harper-Collins.

Shea, G.F. 1994. *Mentoring: helping employees reach their full potential.* New York: American Management Association.

Shonk, J.H. 1997. *Team-based organizations: developing a successful team environment.* Chicago, Ill.: Irwin Professional Publishing.

Singleton, M. 1990. "Programming your subconscious mind for success." *Executive Journal* (June), pp. 8-14.

Smither, L. 2003. "Managing employee life cycles to improve labor retention." *Leadership and Management in Engineering* (Jan.), pp. 19-23.

Snow, C.P. 1965. *The two cultures: a second look.* Cambridge: Cambridge University Press.

Spinner, M.P. 1997. *Project management principles and practices.* Upper Saddle River, N.J.: Prentice Hall.

Stanley, T.J., and W.D. Danko. 1996. *The millionaire next door.* New York: Pocket Books.

Stewart, T.A. 1997. *Intellectual capital: the new wealth of organizations.* New York: Currency/Doubleday.

Strunk, W., Jr., and E.B. White. 1979. *The elements of style,* 3rd Ed. New York: Macmillan.

Tannen, D. 1997. *Talking from 9 to 5: how women's and men's conversational styles affect who gets heard, who gets credit and what gets done at work* (audio cassette). New York: Simon & Schuster Audio.

Telushkin , J. 1996. "Words that hurt, words that heal; how to choose words wisely and well." *Imprimis* (Jan.).

Thompson, J.W. 1996. "Engineers don't always make the best team players." *Electronic Engineering Times* (Sept. 30), p. 124.

Thornberry, N.E. 1989. "Transforming the engineer into a manager: avoiding the Peter Principle." *Civil Engineering Practice* (Fall), pp. 69-74.

Tobias, L.L. 1987. "Hiring for excellence." *Industry Week* (April 20), p. 71.

Townsend, R. 1970. *Up the organization: how to stop the corporation from stifling people and strangling profits.* New York: Alfred A. Knopf.

University of Chicago. 2003. *The Chicago manual of style,* 15th Ed. Chicago, Ill.: University of Chicago Press.

Urban, H. 2003. *Life's greatest lesson: 20 things that matter.* New York: Simon & Schuster.

Walesh, S.G. 1992. "Changing demographics: civil engineering applications." Presented at the 1992 International Convention and Exposition of the American Society of Civil Engineers. New York: ASCE.

———. 1995. "Interaction with the public and government officials in urban water planning." *Hydropolis—the role of water in urban planning: Proceedings of the International UNESCO-IHP Workshop.* Leiden, the Netherlands: Backhuys Publishers.

———. 1996. "It's project management, stupid!" *Journal of Management in Engineering* (Jan./Feb.), pp. 14-17.

———. 1997a. "Job security is an oxymoron." *Civil Engineering* (Feb.), pp. 62-63.

———. 1997b. "More coaching, less osmosis." Editor's Letter. *Journal of Management in Engineering* (July/Aug.).

———. 1999. "DAD is out, POP is in." *Journal of the American Water Resources Association* (June), pp. 535-544.

———. 2000. *Engineering your future: the non-technical side of professional practice in engineering and other technical fields,* 2nd Ed. Reston, Va.: ASCE Press.

———. 2002. *Flying solo: how to start an individual practitioner consulting business.* Valparaiso, Ind.: Hannah Publishing.

Wallem, K., and J. Salimando. 2000. "The liability noose is a bit looser." *American Consulting Engineer* (March/April), pp. 15-18.

Walters, R., and T.H. Kern. 1991. "How to eschew weasel words and other offenses against logic." *Johns Hopkins Magazine* (Dec.), pp. 25-32.

Walther, G.R. 1991. *Power talking: 50 ways to say what you mean and get what you want.* New York: Berkley Books.

Wankat, P.C., and F.S. Oreovicz. 1993. *Teaching engineering.* New York: McGraw-Hill.

Weingardt, R.G. 1997. "Leadership: the world is run by those who show up." *Journal of Management in Engineering* (July/Aug.), pp. 61-66.

———. 2001. "Engineering legends: Octave Chanute and George Ferris." *Leadership and Management in Engineering* (Oct.), pp. 88-90.

———. 2002. "Seeing the forest through the trees." *Structural Engineer* (July), p. 16.

———. 2002. "Engineering legends: Roland C. Rautenstraus and Fred N. Severud." *Leadership and Management in Engineering* (Jan.), pp. 44-56.

Wenk, E., Jr. 1996. "Teaching engineering as a social science." *The Bent of Tau Beta Pi* (Summer), pp. 13-17.

Wiersbe, W.W. 1994. *God isn't in a hurry.* Grand Rapids, Mich.: Baker Books.

Ziglar, Z. 1986. *Top performance.* New York: Berkley Books, pp. 134-139.

Zinsser, W. 1988. *Writing to learn.* New York: HarperResource.

Legal columns

"Legal Affairs Section," in *Journal of Professional Issues in Engineering Education and Practice*, ASCE.

"Legal Corner," by Arthur Schwartz, in *Engineering Times*, NSPE.

"Legal Counsel," by Michael J. Baker, in *Structural Engineer.*

"Risky Business," by John P. Bachner, in *CE News.*

"The Law" and "Court Decisions," in *Civil Engineering*, ASCE.

Bibles

The Holy Bible, RSV. 1952. New York: Harper & Brothers.

Good News for Modern Man: The New Testament in Today's English Version. 1966. New York: American Bible Society.

The Holy Bible, an American Translation. No date. New Haven, Mo.: Leader Publishing Company.

e-newsletters

"A.Word.A.Day." http:www.wordsmith.org/awad/. Wordsmith.Org.

"Abbott's Communication Letter." http://www.abbottletter.com/. Abbott Management Services.

"CCL's e-Newsletter." http://www.ccl.org. Center for Creative Leadership.

"FMI Leadership Group e-News." http://www.fminet.com/lienews/#top. FMI: Management Consultants to the Construction Industry.

"Great Speaking." http://www.antion.com. Antion & Associates.

"LawMarketing Newsletter." http://www.lawmarketing.com/. Larry Bodine.

"Leadership Wired." http://www.injoy.com. John C. Maxwell.

"Making a Life, Making a Living." http://www.makingalife.com. Mark S. Albion.

"Money Matters." http://quicken.com. Quicken/Intuit.

"Personal Achievement Quote of the Day." http://www.topachievement.com/quote.html. Top Achievement.

"Point Lookout." http://www.chacocanyon.com/. Chaco Canyon Consulting.

"Qmail." dleeds@dorothy.leeds.com. Organization Technologies, Inc.

"SalesDoctors Magazine." http://www.salesdoctor.com

"TaxTalk." http://www.nase.org. National Association for the Self-Employed (NASE).

"TelE-Sales Hot Tips of the Week." http://www.businessbyphone.com/backissues.htm. Art Sobczak.

"The Pastor's Coach." www.INJOY.com/PC. Dan Reiland.

"The Professional Speaker." bill@thebrooksgroup.com. Bill Brooks.

"Winner's Circle Daily Email." An e-newsletter from the Pacific Institute (http://mailman.wolfe.net/mailman/listinfo/wcn), April 25, 2003.

"Word of the Day." http://www.dictionary.com/wordoftheday/list/. Lexico Publishing Group.

"Workbiz Report." http://www.wordbiz.com. WordBiz.com

"Working Solo eNews." http://www.workingsolo.com/.

Websites

http://aerodyn.org/People/vonKarman.html. Information about Theodore von Karman.

ABET. No date. www.abet.org. Accreditation Board for Engineering and Technology.

ASCE, Task Committee on the First Professional Degree. 2001. *Engineering the future of civil engineering* (Oct.). Reston, Va.: ASCE (www.asce.org/ raisethebar).

ASCE, Body of Knowledge Committee of the Task Committee on Academic Prerequisites for Professional Practice. 2003. *Civil engineering body of knowledge for the 21st century: preparing the civil engineer for the future.* Reston, Va.: ASCE (www.asce.org/ raisethebar).

http://chem.ch.huji.ac.il/~eugeniik/history/steinmetz.html. Information about Charles P. Steinetz.

Forrest, D.J. 1999. "Employer attitude: the foundation of employee retention." *Keep Employees, Inc.* (Dec.) http://www.keepemployees.com/WhitePapers/attitude.pdf.

KLO People Dynamics. 2002. "How to avoid the Peter Principle when selecting managers." http://pages.prodigy.net/klo_people_dynamics/pp01acec.htm.

McGloin, J.B. 2003. "Symphonies in steel: Bay Bridge and the Golden Gate." Museum of San Francisco website: http://www.sfmuseum.org/hist9/mcgloin.html.

NSPE. January 2003. NSPE Code of Ethics (http://www.nspe.org).

Reiland, Dan. No date. "The Pastor's Coach" (www.INJOY.com).

Staples.com. 2003. "Power-schmoozing your way to the top" (Jan. 11), (http://www003.staples.com/content/Entrepreneur/PowerSchmoozing.asp?&).

"About Teambuilding, Inc." http://www.teambuildinginc.com. Peter Grazier.

"Academy of Management." http://www.aomonline.org/.

"Advanced Public Speaking Institute." http://www.public-speaking.org.

"ae ProNet." http://www.aepronet.org/. Architects Engineers Professional Network.

"AllBusiness." http://www.allbusiness.com/.

"American Management Association." http://www.amanet.org.

"Center for Creative Leadership." http://www.ccl.org.

"ClassesUSA." http://www.classesusa.com/.

"Free Agent Nation." http://www.freeagentnation.com. Daniel H. Pink.

"International Mentoring Association." http://www.wmich.edu/conferences/mentoring/genrinf.html.

"Kiplinger.com." http://www.kiplinger.com. The Kiplinger Washington Editors, Inc.

"Learnon.org." http://learnon.org. American Society for Engineering Education.

"Livelink Virtualteams." htttp://www.virtualteams.com/index.asp. NetAge.

"Making a Life, Making a Living." http://www.makingalife.com/. Mark S. Albion.

"Management First." http://managementfirst.com/.

"Michael Greer's Project Management Resources." http://www.michaelgreer.com/.

"MSN Money." (http://moneycentral.msn.com/home.asp). CNBC.

"National Society of Professional Engineers." http://www.nspe.org.

"National Speakers Association." http://www.nsaspeaker.org/.

"Nolo Law for All." http://www.nolo.com. Nolo.

"Organisational Change." http://organisationalchange.co.uk/.

"Presentation Skills, Public Speaking and Professional Speaking." http://www.antion.com. Antion and Associates.

"Quicken." http://www.quicken.com. Intuit.

"SBA Starting Your Business." http://www.sba.gov/starting. U.S. Small Business Administration.

"Smartmoney." http://www.smartmoney.com/pf/. *Smart Money* magazine.

"Team Management Systems." http://www.tms.com.au. Australian Consultants.

"The Project Management Institute." http://www.pmi.org/.

"Toastmasters International." http://www.toastmasters.org/.

"Top Achievement." http://www.topachievement.com/.

"Virtual Projects." http://www.vrtprj.com. Rainer Volz.

"Working Solo." http://www.workingsolo.com. Terri Lonier, SOHO (Small Office/Home Office).

Index